Ignition, Valve Timing

and

Automobile Electric Systems

(SELF-STARTING AND LIGHTING)

A Comprehensive Manual of Self-Instruction on the Operation, Adjustment and Repair of Magnetos, Battery Ignition Systems, and Self-Starting Mechanisms. Complete Tables and Data on Valve Timing for a Great Number of American Automobiles

The Ford Ignition System and Its Adjustment

By JOHN B. RATHBUN

Formerly Editor of "Ignition and Accessories." Author of "Gas Engine Troubles and Installation," "Gas, Gasoline and Oil Engines," "Aeroplane Engines," and "Aeroplane Construction and Operation."

TROUBLE CHART

FOR

IGNITION AND STARTING SYSTEM

British Library Cataloguing-in-Publication Data
A catalogue record for this book is available from
the British Library

A History of Electrical Engineering

Electrical engineering is a field of engineering that generally deals with the study and application of electricity, electronics, and electromagnetism. The field first became an identifiable occupation in the latter half of the nineteenth century after commercialization of the electric telegraph, the telephone, and electric power distribution and use. Subsequently, broadcasting and recording media made electronics part of daily life – the effects of which we are only just beginning to understand in the present day.

Electricity has been a subject of scientific interest since at least the early seventeenth century. The first electrical engineer was probably William Gilbert who designed the versorium: a device that detected the presence of statically charged objects. He was also the first to draw a clear distinction between magnetism and static electricity and is credited with establishing the term electricity. Despite this, it was not until the nineteenth century that research into the subject started to intensify. Notable developments in this century include the work of Georg Ohm, who in 1827 quantified the relationship between the electric current and potential difference in a conductor, Michael Faraday, the discoverer of electromagnetic induction in 1831, and James Clerk Maxwell, who in 1873 published a unified theory of electricity and magnetism in his treatise *Electricity and Magnetism*.

Beginning in the 1830s, efforts were made to apply electricity to practical use in the telegraph. By the end of the nineteenth century the world had been forever changed by the rapid communication made possible by engineering development of land-lines, submarine cables, and, from about 1890, wireless telegraphy. Practical applications and advances in such fields created an increasing need for standardized units of measure. They led to the international standardization of the units 'volt', 'ampere', 'coulomb', 'ohm', 'farad', and 'henry' – achieved at an international conference in Chicago 1893.

During these years, the study of electricity was largely considered to be a subfield of physics. It was not until about 1885 that universities and institutes of technology such as Massachusetts Institute of Technology (MIT) and Cornell University started to offer bachelor's degrees in electrical engineering. The Darmstadt University of Technology founded the first department of electrical engineering in the world in 1882, and in 1885 the University College London founded the first chair of electrical engineering in Great Britain.

During these decades, use of electrical engineering increased dramatically. In 1882, Thomas Edison switched on the world's first large-scale electric power network that provided 110 volts – direct current (DC) – to fifty-nine customers on Manhattan Island in New York City. In 1884, Sir Charles Parsons invented the steam turbine allowing for more efficient electric power generation. Alternating current, with its ability to transmit power more efficiently over long distances via the use of transformers power system developed rapidly in the 1880s and 1890s. The spread in the use of AC set off in the United States what has been called the *War of*

Currents between a George Westinghouse backed AC system and a Thomas Edison backed DC power system, with AC being adopted as the overall standard.

Modern developments to the field of electrical engineering have since been made at an astounding rate. During the development of radio, many scientists and inventors contributed to radio technology and electronics. The mathematical work of James Clerk Maxwell during the 1850s had shown the relationship of different forms of electromagnetic radiation including the possibility of invisible airborne waves (later called 'radio waves'). In his classic physics experiments of 1888, Heinrich Hertz proved Maxwell's theory by transmitting radio waves with a spark-gap transmitter, and detected them by using simple electrical devices. In 1895, Guglielmo Marconi followed up on this work, and adapted the known methods of transmitting and detecting these 'Hertzian waves' into a purpose built commercial wireless telegraphic system. He eventually managed to transmit the wireless signals across the Atlantic between Poldhu, Cornwall, and St. John's, Newfoundland – a distance of 3,400 kilometres.

From this point onwards, developments in electrical engineering came rapidly. In 1920, Albert Hull developed the magnetron which would eventually lead to the development of the microwave oven (in 1946 by Percy Spencer). In 1934 the British military began to make strides toward radar (which also uses the magnetron) under the direction of Dr Wimperis, culminating in the operation of the first radar station at Bawdsey in August 1936. In 1941, Konrad Zuse presented the Z3, the world's first fully functional and programmable computer using electromechanical parts. The arithmetic performance of these new machines allowed engineers to develop completely new technologies and achieve novel objectives, including the Apollo program which culminated in landing astronauts on the Moon.

Today, electrical engineering has subdivided into a wide range of subfields including electronics, digital computers, power engineering, telecommunications, control systems, radio-frequency engineering, signal processing, instrumentation, and microelectronics. It has helped to develop a massive array of technology, used all over the world – and as a discipline, continues to develop in the present day. It is hoped that the current reader enjoys this book on the subject.

ELECTRICAL SYSTEM OF A MODERN AUTOMOBILE

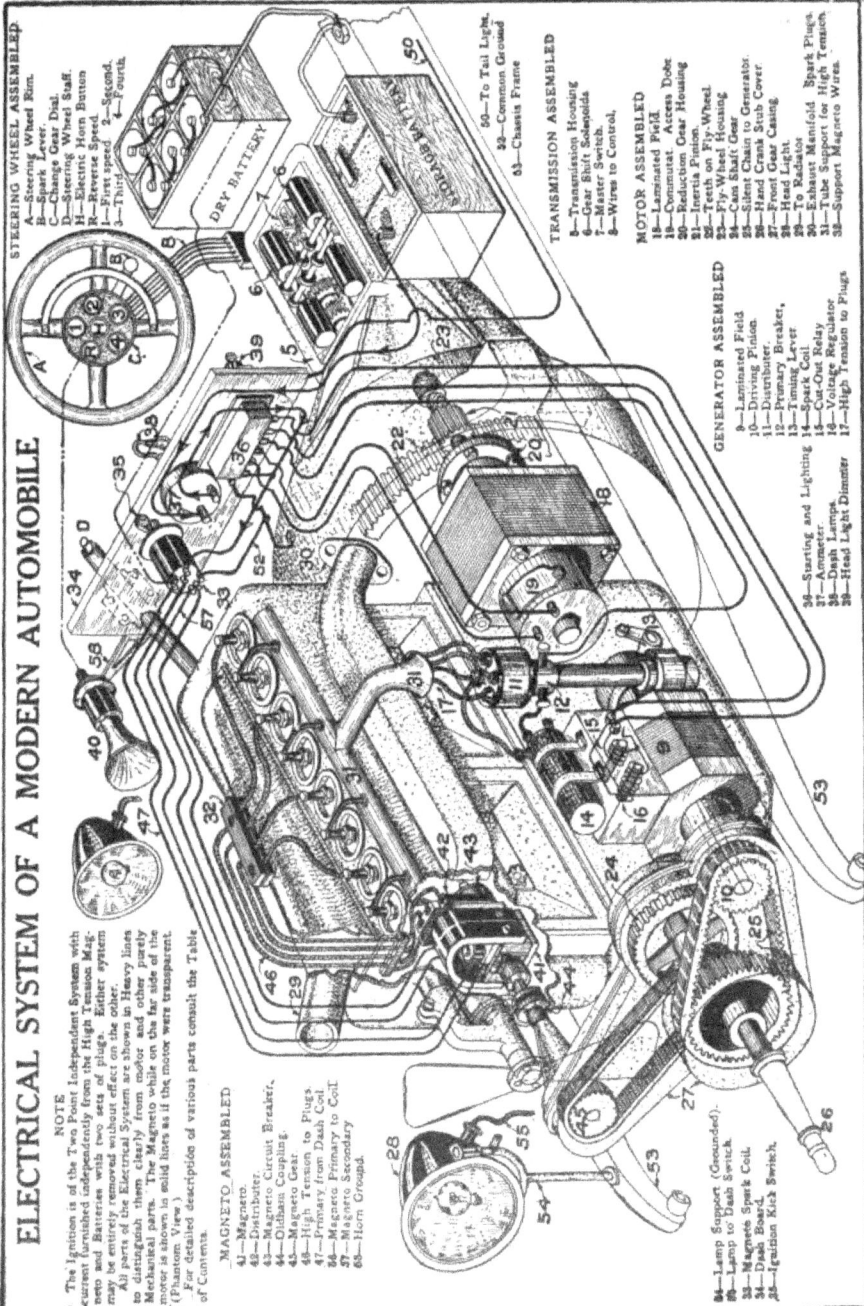

NOTE

The Ignition is of the Two Point Independent System with current furnished independently from the High Tension Magneto and Batteries with two sets of plugs. Either system may be entirely removed without effect on the other.

All parts of the Electrical System are shown in Heavy lines to distinguish them clearly from Mechanical parts. The Magneto while on the far side of the motor is shown in solid lines as if the motor were transparent. (Phantom View.)

For detailed description of various parts consult the Table of Contents.

MAGNETO ASSEMBLED

41—Magneto.
42—Distributor.
43—Magneto Circuit Breaker.
44—Offbase Coupling.
45—Magneto Gear.
46—High Tension to Plugs
47—Primary from Dash Coil
56—Magneto Primary to Coil
57—Magneto Secondary
58—Horn Ground.

54—Lamp Support (Grounded).
55—Lamp to Dash Switch
53—Magneto Spark Coil.
52—Dash Board.
51—Ignition Kick Switch.

STEERING WHEEL ASSEMBLED

A—Steering Wheel Rim.
B—Spark Lever.
C—Change Gear Dial.
D—Steering Wheel Staff.
E—Electric Horn Button
R—Reverse Speed
1—First speed 2—Second.
3—Third 4—Fourth.

50—To Tail Light.
49—Common Ground
48—Chassis Frame

TRANSMISSION ASSEMBLED

5—Transmission Housing
6—Gear Shift Solenoids.
7—Master Switch.
8—Wires to Control.

MOTOR ASSEMBLED

18—Laminated Field
19—Commutat. Access Door.
20—Reduction Gear Housing
21—Inertia Pinion.
22—Teeth on Fly-Wheel.
23—Fly-Wheel Housing
24—Cam Shaft Gear
25—Silent Chain to Generator.
26—Hand Crank Stub Cover.
27—Front Gear Casing.
28—Head Light.
29—To Radio
30—Exhaust Manifold Spark Plugs.
31—Tube Support for High Tension.
32—Support Magneto Wires.

GENERATOR ASSEMBLED

9—Laminated Field
10—Driving Pinion.
11—Distributor.
12—Primary Breaker,
13—Timing Lever
14—Spark Coil
15—Cut-Out Relay
16—Voltage Regulator
17—High Tension to Plugs

34—Starting and Lighting
37—Ammeter.
35—Dash Lamps.
39—Head Light Dimmer

DRY BATTERY

Preface

Many of the troubles from which motorists have suffered in the past—and still suffer, in fact, despite recent improvements in construction of all the essential parts of the automobile—have arisen from failure of the ignition system to perform its proper function. While these troubles may perhaps be minimized in the latest model cars, there are still in daily use in the United States and Canada many thousands of machines built and equipped in the days of motor-car development, and to every owner and operator, no matter whether his car be new or old, the subject of ignition is of the utmost importance.

To know what to do in case of ignition troubles, it is imperative to learn something definite about the principles of the ignition system used on the car. Intelligent handling of the car in emergencies can only be assured when the operator possesses such information. It will not pay to "go it blind" in seeking the causes of ignition failure. When the engine stops or misbehaves from such causes knowledge is indeed "power."

The object of this treatise is to equip the reader with such a knowledge of the interesting subject of Ignition that he will be able to handle his own particular apparatus with intelligence and skill. The mere consciousness that he understands the principles and construction of his ignition devices will add immensely to his comfort on the road, giving him greater confidence in himself as a driver and stripping the ignition bogey of most of its terrors.

Every modern ignition device, whether for the automobile, motor-boat or stationary engine is described so fully and at the same time in so simple a manner that the text is of equal value to the beginner and the veteran repair man. No previous knowledge of electricity is necessary for a full understanding of the matter contained in this volume. Examples of well known commercial apparatus follow the descriptions of the various systems, together with notes on their installation and repair. Particular attention is paid to the operation of the high tension magneto and to the various forms of dual ignition circuits.

Probably the most unique and useful features of the book is the chapter on valve timing, a subject on which very little has appeared heretofore in book form. Even the little that has been published has been so elementary or so indefinite that it was practically useless to the practical man. In contrast we publish the exact timing figures issued by nearly every prominent automobile manufacturer for different models of cars so that even a beginner can obtain the correct results without experiment. Firing orders of four, six, seven, eight and twelve cylinder motors are given in addition to the material on timing.

As the electric self starting and lighting system bears a definite relation to the ignition system on most modern cars we have covered the construction, operation and repair of this feature in a very simple and careful manner. A complete chapter is devoted to the care of the ignition and starting storage battery alone.

THE AUTHOR.

CONTENTS

CONTENTS

CHAPTER I.

IGNITION PRINCIPLES.

Combustion. The gasoline and gas engine are of a class of prime movers known as "Heat engines"—that is, they convert heat energy into useful mechanical energy through a process of combustion. The mixture of fuel and air is burned within the cylinder, after it has been compressed to a comparatively high pressure, and the sudden increase in pressure due to the expansion causes the piston to move against the load and deliver energy to the engine shaft. The fuel must always be vaporized or in a gaseous state when used in a gas or gasoline engine, and must be mixed with the proper proportion of air in order to burn it to its lowest chemical condition. It will be seen that the ignition or "Kindling of the mixture is of the greatest importance, and the various methods of obtaining the igniting spark and the proper regulation and adjustment of the different parts of the apparatus are vital elements in the operation of a gas or gasoline engine.

In any engine, the mixture of fuel and air must be highly compressed in order to obtain efficient combustion, and a reasonable amount of work out of a given size cylinder. If the gas is ignited at atmospheric pressure, the combustion is slow, much heat is radiated and lost, and the resulting pressure due to the expansion of the gas is relatively low. Thus an engine compressing to 60 pounds per square inch will only develop about one horsepower for every 10 cubic inches displaced by the piston, while compressing to 110 pounds per square inch will increase the capacity to a point where 3.8 cubic inches will produce one horsepower. The thermal efficiency will be increased from about 18 percent to 28 percent.

Engine Cycles. In order to operate continuously and deliver power, the engine must go through a routine of operations, each act being performed over and over in the same order. Each of these operations is known as an "Event" and the complete series of events.is.known as a "Cycle" or as a "Cycle of Events." In any gas engine the following events must take place: (1) The cylinder is filled with an explosive mixture, (2) The mixture is compressed, (3) Ignition takes place, (4) The piston moves out on the working stroke to produce the power (5) The exhaust gases are liberated and then cleaned out of the cylinder. The engine is classified by the number of piston strokes taken to accomplish this cycle of events, for there are a great number of possible combinations between the events and the number of strokes required for the cycle. Thus, a two-stroke cycle motor completes the five events in two strokes, or one revolution. A four-stroke cycle engine goes through the series once in every four strokes, or in every two revolutions. As the four-stroke cycle is used almost exclusively in automobile practice, we will confine ourselves to a description of this type.

A full understanding of the cycles is of the utmost importance in ignition and valve setting, and the subject must be thoroughly studied and committed to memory before proceeding further. The opening and closing of the valves, and the timing of the ignition apparatus, of course depends entirely upon the time at which the events take place in regard to the piston position.

Four-Stroke Cycle. This cycle, which is often called the "Four-Cycle," is the most commonly used for stationary and automobile engines. As before explained, the five events take place during four strokes, or two revolutions of the crankshaft. According to the strokes, the events take place in the following order:

Stroke 1 (Suction Stroke). The piston moves toward the crank and sucks in a charge of the combustible mixture.

Stroke 2 (Compression Stroke). The piston now moves back in the opposite direction, compressing the mixture. At the end of the compression stroke, the spark occurs and fires the gas.

Stroke 3 (Working or Power Stroke). The hot, expanded gas creates a high pressure on the piston, and produces the power. At the end of the stroke, the pressure is much reduced, the exhaust valve opens, and the gas rushes out to atmosphere.

Stroke 4 (Scavenging Stroke). The piston now returns, moving away from the crank, and pushes out the residual gas left in the cylinder.

At the end of the fourth stroke, the cycle starts over again by sucking in a charge for the next combustion. It will be noted that two events occur at the end of the compression stroke (Stroke 2)—that is, both ignition and compression. In practice, the spark does not occur at the exact end of the compression stroke, but a few degrees before, as the gas takes time to burn, and combustion is assumed to be completed at the center. Since the exhaust gas takes a definite length of time in which to get out of the cylinder, and thus reduces the pressure to that of the atmosphere, it is allowed to start before the piston actually reaches the end of the stroke.

Fig. 1 is a diagrammatic outline of the four-stroke engine with the piston (P) starting out on the suction stroke, the piston moving down along the arrow (m). Through the connecting rod (R), the piston is moving the crank (C) in the direction of the arrow (n). The combustible mixture formed in the carbureter is drawn into the cylinder through the inlet port (I) and the open inlet valve (B). This valve is opened and closed by the cam (IC), which is driven from the crankshaft through the gear (G'). The exhaust valve (A), which allows the burnt gas to escape to atmosphere, remains closed during the suction stroke, as shown. Slightly after the piston reaches the lower end of the suction stroke, and starts up on the compression stroke, the inlet valve (B) closes.

Fig. 2 shows the piston traveling up on the compression stroke in the direction of the arrow (m'), and as will be seen, both the valves (A) and (B) are closed so that pressure can be built up. At the upper end of the compression stroke, the spark (S) occurs in the spark plug (Q), and in such a way that the spark is in direct contact with the mixture and thus starts combustion. Theoretically, combustion should be complete before the piston starts moving down on the working stroke, but practically the spark occurs slightly before this point, so that the flame will have time to spread through the entire volume of the mixture. The crank is still revolving in the direction of the arrow (n), and after the ignition takes place the piston starts down on the working stroke.

Fig. 3 shows the piston near the end of the working stroke, and at this point the pressure is reduced and the exhaust valve (A) is opened for the escape of the burned gas, as shown. This valve movement is produced by the exhaust cam (EC), driven from the crankshaft through the timing gear (G). In the majority of engines, the exhaust valve opens from 40 to 45 degrees of crank travel before the crank reaches the lower dead center. Crank rotation still in the direction of (n).

Fig. 4 shows the exhaust valve still open on the "Scavenging" return stroke, during which time the piston is pushing out the remainder of the gas. Near the upper end of this stroke, the exhaust closes, and the inlet opens ready for the next suction stroke and the next cycle. The

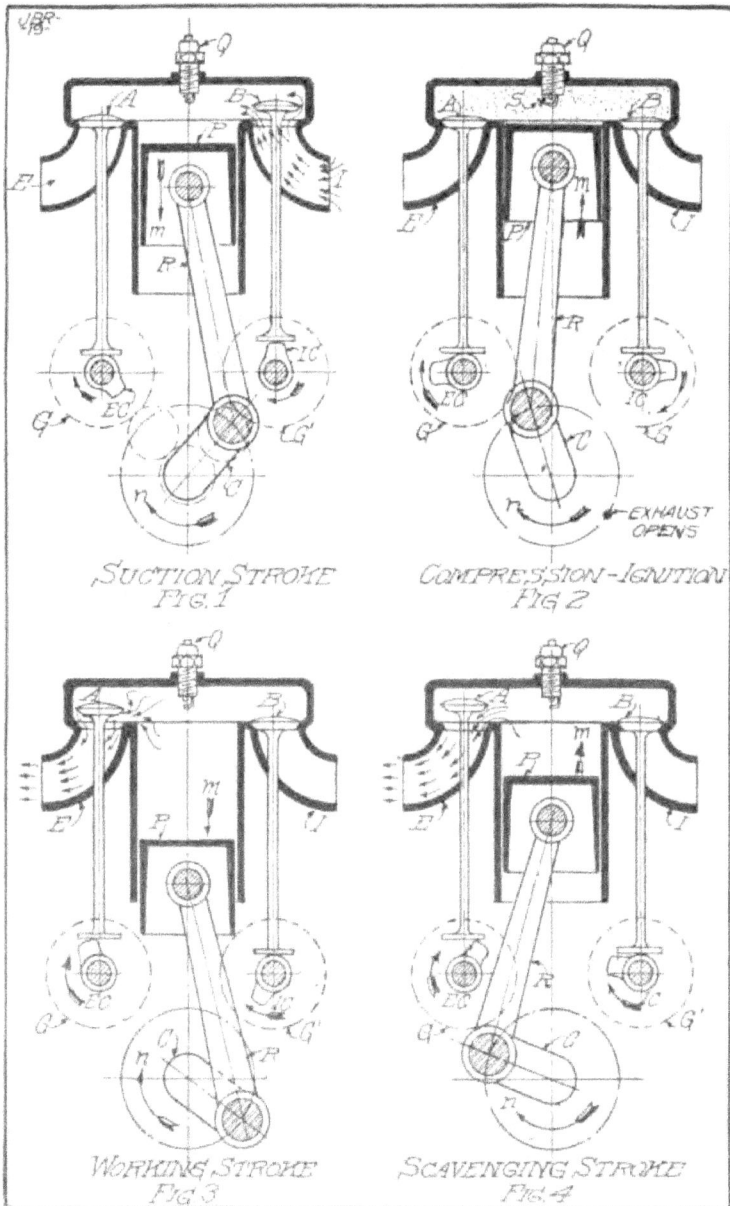

Figs. 1-2-3-4. The Five Events of the Four Stroke Cycle Engine.

burnt gas retained in the cylinder during this stroke is reduced to atmospheric pressure, but even this must be expelled in order to prevent the dilution and contamination of the next charge of mixture. A very small percentage of burnt gas will seriously reduce the capacity of the engine.

The engine makes two revolutions during this cycle, and as the valves only operate once during this time, it is evident that the cams (EC) and (IC) must operate at exactly one-half the crankshaft speed. For this reason, the ratio of the gears (G) and (G') to the driving gear on the crankshaft is as one is to two, and are therefore often referred to as the "Half-time Gears." The exhaust cam (EC), and the inlet cam (IC), are of different form, as the inlet and exhaust valves remain open for different periods of time. The spark is controlled by an automatic switch known as a "Timer" or "Breaker," and is also driven at some exact relation to the camshaft speed. The exact speed ratio of the spark timer depends a great deal upon the number of cylinders used, hence this is likely to vary from 1 to 1, up to 4 to 1.

Ignition Systems. Up to the present, we have only considered the spark as a spark without inquiring into its origin. There are many ways of producing an igniting spark or flame, and during the development of the internal combustion engine the following methods have been employed: (1) OPEN FLAME ignition, in which the flame ignited the gas through a "Touch hole" after the manner of an old-fashioned cannon; (2) HOT TUBE ignition, in which the gas was ignited by contact with the incandescent walls of a tube; (3) HOT WIRE ignition, in which an electrically heated wire fired the charge; (4) CATALYTIC ignition, produced by the condensing effect of platinum black; (5) COMPRESSION, as in the Diesel engine, by which a spray of fuel is vaporized and ignited by coming into contact with highly compressed air; and

(6) ELECTRIC ignition, performed by producing an electric spark in the midst of the mixture. Ignition by methods (5) and (6) are by far the most common, and it is very seldom that one encounters any of the remaining systems. On gas and gasoline engines, electric ignition is used almost exclusively, but the compression system is especially adapted to heavy oil and kerosene engines, and for these fuels has almost entirely supplanted electricity.

Hot Tube Ignition. In the older engines the hot tube was very commonly used, but it had many drawbacks and was finally superseded by the electric system. Fig. 5 shows the cylinder (C), with the hot tube (B) screwed into one end of the cylinder. The tube is opened into the cylinder, but is closed at the outer end, and is heated to incandescence by the burner flame (A). When the piston moves up on the compression stroke, the mixture is forced back into the tube until it finally comes into contact with the upper hot portions of the tube, and ignition occurs. By arranging the length of the heated portions of the tube or by moving the flame, the time of ignition can be controlled to some extent. The further up the hot portion was located, the longer it took for the gas to reach the heat, and the later the ignition occurred. This was decidedly unsatisfactory, for the tubes would burn out, and it was difficult to exactly control the timing. With the engine throttled down on light loads, the compression would not be sufficient to force the gas back to the hot spot, and the engine was likely to misfire. For this reason, it was practically restricted to hit-and-miss engines.

Electric Ignition. The electric system has succeeded the hot tube and presents many advantages, although it is still far from being an ideal ignition method. It depends for its operation upon the fact that a current of electricity passing through an air gap heats the air to incandescence, and the heat thus produced will ignite any mixture in the vicinity. The resistance offered by the air, when current is jumping or flowing between two points, generates a heat in the same way that heat is produced by resisting the pull of a rope, or in fact by resisting any force in motion. The more rapid the flow of current, and the greater the resistance offered, the higher will be the temperature. Since the resistance of air is many thousands of times greater than the resistance of the copper wire

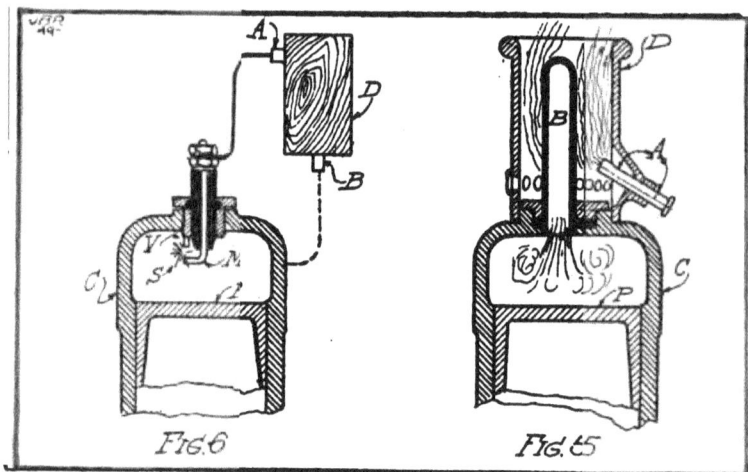

Figs. 5 and 6. The Two Principal Forms of Ignition. Fig. 5 Is the Old Hot Tube, While Fig. 6 Is the Jump Spark System.

through which it flows, it is evident that a given current will produce much more heat in the air gap than in the conducting wire, especially with the air gap concentrated in a very small space.

Fig. 6 shows an electric spark gap (S) installed in the cylinder (C). The current is produced by some source which we will indicate by (D), and is conducted to the spark gap by the wires (A) and (B). We will assume the current to flow from (D) through the wire (A), and then down the center electrode or conductor (M). At the lower end of the electrode (M), is a small air gap (S), which separates the electrode from the opposite point (V). After jumping across (S), between (M) and (V) the current flows back to the source through the cylinder walls (C) and the return wire (B), which is shown dotted. The conductor (B) is not always a wire, but in many cases is formed by the metal frame of the engine. This serves the same purpose and avoids much complication and many troublesome connections. The path through which the current makes its round trip from the source, and then back to the source (D), is called a "Circuit," and this must be complete from start to finish before the current will flow.

Chemical Batteries. In the early ignition systems, the current was nearly always produced by batteries, and this practice still exists to a limited extent. The primary battery produces current by chemical means, the chemical reaction consuming the battery elements in exactly the same way that heat energy is produced by the combustion of coal. In the majority of primary batteries, zinc is used as the fuel, and this is converted into zinc chloride with the further loss of the acid or corrosive fluid. As this method of producing current is simple, inexpensive, and has not moving parts, it is very satisfactory for some branches of ignition service.

A dry cell and its circuit is shown by Fig. 7. This consists of a zinc can (C), in which is a paste of sal ammoniac solution and some absorbent material such as blotting paper. In the center of the can, and in contact with the paste is the carbon rod (A), which is also known as the "Positive Electrode." The action of the sal ammoniac (Electrolyte) on the zinc and carbon surfaces establishes an electrical difference of potential, or difference in electrical pressure, which causes the current to flow around the circuit. In the figure, the carbon rod (A) and the zinc can (B) are connected with the copper wire (W), thus forming a circuit through which the current flows in the direction shown by the arrows. It will be noted that the current flows from the carbon or from the "positive" connection to the Zinc or negative connection. In the industry, the positive pole is designated by (+), and the negative by (–), as marked in the figure. On the inside of the battery cell, the current flows through the paste in the opposite direction, or from negative to positive, but this latter fact need not interest us much at this point. The direction of flow is important, for in some appliances it is necessary to maintain the flow in a certain direction for the operation of the device. A current which flows continuously in one direction, as shown, is called a "Continuous Current" or a "Direct Current."

Magnetic Effect of Current. When an iron bar is surrounded by a coil of wire, as shown by Fig. 8, and current is flowing through the coil, the bar is magnetized and becomes capable of attracting and holding pieces of iron or steel as long as the current flows. If the bar is of very soft iron or steel, it demagnetizes instantly, and drops the pieces that it had been holding. In Fig. 8 the magnetizing coil, or "Solenoid," is indicated by (C), and the iron bar by (S-N), the arrows showing the direction of current flow around the iron "Core." The current is supplied by the dry cell at the left. The effect produced by the coil is that one end of the bar is magnetized to a North seeking polarity (N), while the other end (S) would point to the South pole of the earth if the bar and coil were pivoted and allowed to swing like the needle of a compass. This tendency to turn in a fixed relation to the earth is a distinguishing property of magnets, and holds true of electromagnets as well as with the permanent magnets more commonly used in compasses. A field of magnetic force extends around the bar (Magnetic Field), and any metallic body lying in the field is influenced by it. The general outline of the magnetic field is indicated by the dotted lines (M). There are only a few substances that are affected by the magnetism or which can be magnetized. These are iron, steel, nickel, and one or two very rare metals. Brass or wood are not magnetized, but they do not prevent the passage of the electrical field—in fact, all non-magnetic substances are practically transparent to the magnetic force.

The external magnetic effect is always greatest at the poles (S) and (N), the attractive power growing less and less as we approach the center of the bar. It should also

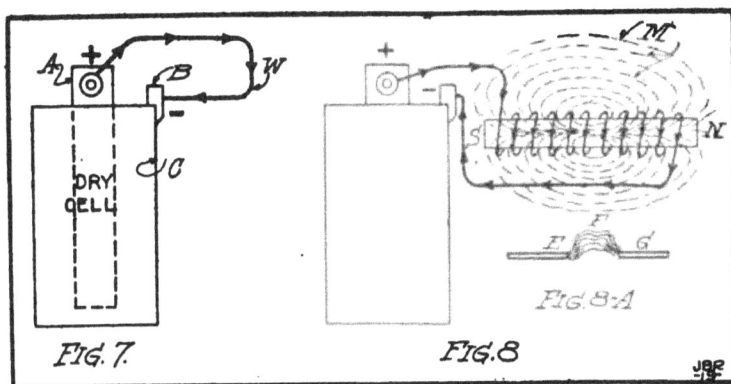

Figs. 7-8. Simple Electric Circuits.

be noted that two poles, North and South, are always produced, and that neither pole can exist alone. The polarity of the magnetic bar bears a fixed relation to the direction in which the current flows through the coil (C). If the fingers are pointed in the direction of current flow and the thumb is held at right angles to the fingers, then the thumb will point to the North pole of the bar. When the current starts to flow and build up in strength, the magnetism increases with it and the Magnetic Field (M) spreads out from the bar. When the current is cut off, the field immediately contracts and returns toward the bar, and cuts or travels through the turns of the coil. Since it is the common practice to illustrate magnetic force by a series of lines, as shown, the magnetic force is sometimes called "Lines of Force" or "Magnetic Lines," and in some cases it is called the "Magnetic Flux."

The current flow cannot build up instantly in a circuit containing a coil, and the value of the electric current builds up very slowly when the number of turns or the quantity of iron within the coil is increased. The retarding influence is known as "Inductance," and this is caused by an opposing current produced by the spreading magnetic lines of force acting on the turns of the coil. When magnetic lines of force cut through conductors or wires, they generate an electric current in the wire, and this always opposes the flow of the current that energizes the bar. When the electric circuit is broken, the contracting lines of force cut through the coil and generate a current that tends to continue the current. This "Induced" current is in evidence when a circuit is broken, the induced current appearing as a flame that continues for a short time at the point where the wires are separated. The more the turns, and the greater the mass of iron in the core (S-N), the greater will be the induced current voltages. Thus in Fig. 8-a, the separation of the wire ends (E) and (G) produces a gap which the voltage of the induced current is able to break down and thus produce the arc or flame (F). A straight, simple circuit has very little inductance.

Make-and-Break System. In the make-and-break system advantage is taken of the inductance, and the terminals of a highly inductive circuit are broken or separated in the combustion chamber of the engine, thus producing an igniting spark when the circuit is opened. The coil (Spark coil) consists of a number of turns of heavy copper wire, while the core (S-N) is built up of the very fine soft-iron wire which will magnetize and demagnetize very rapidly when the circuit is closed and opened. This method is generally applied to large, slow-speed engines, for the weight and inertia of the moving parts prevents effective operation at very high speeds.

Series Battery Connections. The electrical pressure or "Voltage" developed by a single dry cell is very low, about 1.5 volts as an average. As the voltage required by the ignition system ranges from 6 to 8 volts, we must find some means of increasing the voltage. This is done by connecting a number of cells in series, as shown in Fig. 9, the total voltage of the three cells 1, 2 and 3, being

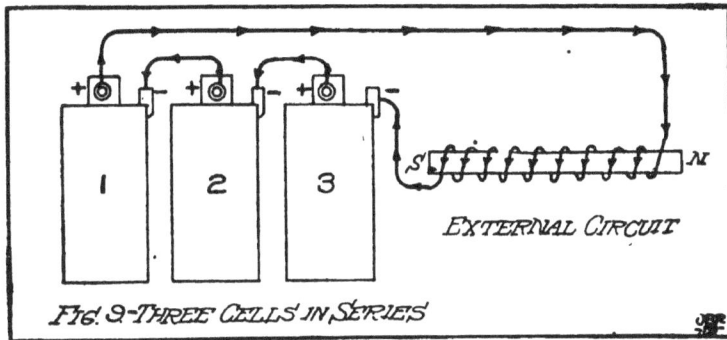

Fig. 9—Three Cells in Series

equal to the sums of the cell voltages, or 1.5+ 1.5+ 1.5 = 4.5 volts. For 6 volts we would use 6.0/1.5 = 4 cells.

It will be noted that in series connection the positive connection post of one battery cell is connected to the negative of the next cell, or in other words, the connection posts of opposite polarity are connected. The external circuit is then connected to the unoccupied terminals of the two end cells, as shown, the external circuit in this case being the coil or solenoid (S-N).

Induction or Secondary Coil. The voltage required for forcing the igniting spark across the points of the spark plug is very high, this requiring many thousand volts. It would be impractical to obtain this voltage directly by batteries, for then a 15,000-volt ignition circuit would require 10,000 dry cells connected in series. The use of an inductance or "Booster" coil in the circuit, as already explained, is not possible in many cases, so that we must have recourse to some magnetic system by which we can boost up the low battery voltage to the high potentials required at the spark gap. This is accomplished by a double-wound coil, known as an induction or secondary coil (Transformer Coil), which magnetically increases the voltage in the battery coil to a high voltage in the secondary coil. This is shown by Fig. 10.

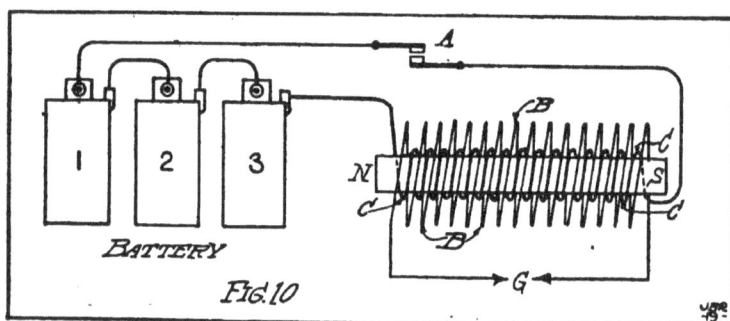

Fig. 10

A soft-iron core (N-S) is wound with a few turns of heavy wire (C), and is called the primary coil. It is through this coil that the battery current flows, and hence is the coil which magnetizes the core (N-S). Over the primary coil, and thoroughly insulated from it, is the secondary coil (B), which consists of many thousands of turns of very fine wire. The primary coil is connected to the batteries 1, 2 and 3, so that the primary circuit can be closed and opened by the switch (A). The ends of the secondary winding are separated by the spark gap (G), and this corresponds to the spark gap of the plugs.

When the primary switch (A) is closed, the battery current flows through the primary coil and magnetizes the core (N-S). The expanding magnetic flux cuts through the turns of the secondary coil and thus generates a current in the secondary. Since the voltage induced is proportional to the number of turns, the thousands of turns in the secondary are capable of being arranged for almost any voltage desired. The ratio of the secondary and primary voltage is roughly proportional to the ratio of turns of the secondary to the turns in the primary. Thus, with 36,000 turns in the secondary, and 40 turns in the primary, the voltage in the secondary will be approximately: 36,000/40 = 900 times. If the battery voltage in the primary is 6 volts, then the secondary voltage will be: 6 × 900 = 5,400 volts.

It should be noted that the current in the secondary is "Alternating"—that is, the current flows back and forth instead of in the same direction as with the continuous battery current. The secondary flows in one direction when the primary switch is closed, and in the opposite direction when the primary is opened. The voltage is greater when the switch is opened and the primary circuit is broken, for this causes the most abrupt change in the speed of the magnetic flux, and hence the highest voltage.

High Tension Ignition Circuit. Several modifications must be introduced into the circuit of the simple secondary coil, just shown. The switch in the primary circuit must be opened automatically at the instant that the piston arrives at the end of the compression stroke, so that ignition will take place here. Again, since the change in the value of the magnetic flux in the core must be as quick as possible, we must provide some device in the primary circuit that will absorb the inductive spark that occurs when the switch is opened. This spark causes the current to flow for an appreciable time after the switch is opened, thus prolongs the duration of the primary flow and causes the voltage to increase very slowly. If a condenser is connected across the terminals of the primary coil, or across the contact points, it will absorb the inductive energy developed in the primary, and thus will greatly increase the "Snap" and energy of the secondary spark in the plugs.

Fig. 11 shows the elementary features of a high-tension ignition circuit. A revolving switch or "Timer" is driven at a given speed by the engine, and alternately opens and closes the primary coil (C). The switch blade is so located on its shaft that it breaks the primary circuit just before the piston reaches the end of the compression stroke. The revolving switch arm is (A), the shaft is (B), and the stationary contact segment is (D). A condenser (I) is connected across the ends of the primary coil to suppress the sparking at the breaker points (D) and (R), and to increase the energy of the secondary circuit.

One end of the secondary coil (S) is connected with the spark plug (F) inserted in the cylinder (J). The high-tension current passes through the plug and into the combustion chamber, causing a spark to take place at the spark gap (H). The wire is well insulated from the metal of the cylinder, and after the spark jumps the gap the current returns to the secondary winding through the frame of the engine, as indicated by the dotted line (E) and (E'). This circuit contains the essential elements of the high-tension battery ignition system.

The Condenser. Fig. 12 shows a detail of the condenser and its construction. The primary coil is shown by (C) connections to the condenser (D), serving to lead the induced current to the condenser. The condenser proper is made up of alternate piles of tinfoil and paper, the paper serving to insulate the adjacent tinfoil leaves. One-half the tinfoil sheets are connected to one end of the coil by (A), while the other half of the sheets are connected to the coil terminal (B). In the figure the tinfoil sheets (t-t-t-t) are connected to (A), while the sheets (t'-t'-t') are connected with terminal (B), the plates being piled alternately right and left. In very good coils the

condenser insulation is mica instead of paper, and this
gives excellent results but is expensive.

Firing Multiple Cylinders. When more than one cylinder is used, a separate coil may be used for each cylinder,
or a single coil may be provided with a revolving "Distributer" switch, which connects the cylinders with the
coil alternately and in the proper firing order. The Ford
automobile uses four coils, one for each cylinder, but the
majority of cars use a single coil with a high-tension dis-

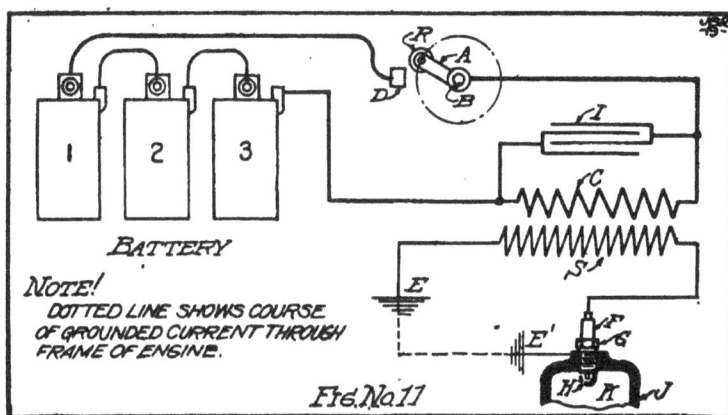

Fig. 11.   High Tension Circuit.

tributer. In any case the primary circuit is interrupted
periodically by the "Timer," so that each coil receives
its proper primary impulses at the correct time. The
timer revolves at camshaft speed, or at half the entire
revolutions.

Storage Batteries. When a dry cell is exhausted or
"used up," it cannot successfully be restored to its former
usefulness, and it is cheapest to throw it away. The zinc
is badly eaten away, and the solution or "Electrolyte" is
neutralized and practically destroyed. By passing current backwards through the cell, some of the zinc is taken
out of the solution and deposited on the zinc, but this

method takes considerable time and patience and does not
turn out right. With wet batteries the solution and zinc
element can be renewed easily and cheaply, and the wet
cell is generally best for stationary engines. For automobiles and portable engines wet batteries are not a
decided success because of the slopping of the solution
and the chances of leakage.

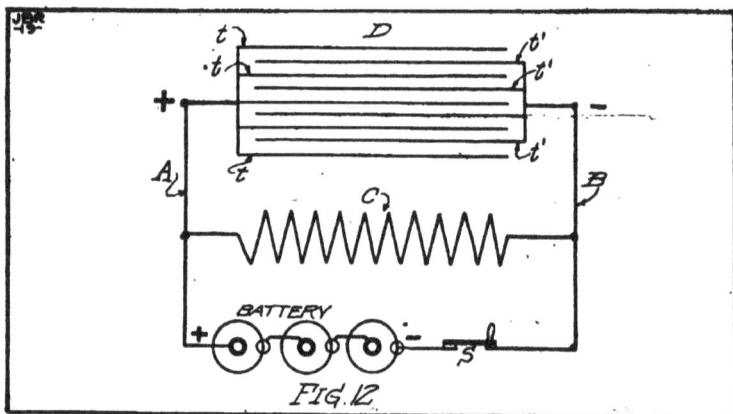

Fig. 12. Primary Circuit.

The storage or "Secondary" battery is entirely different
in construction from the primary battery, and can be
recharged repeatedly by passing a charging current
through the cell in a direction opposite to that furnished
by the battery. This is a very convenient system, for the
battery will absorb electrical energy while the engine is
running, and will give it up either while running or when
the engine is stopped. The voltage of a storage battery
is higher than that of a primary cell, averaging from 2.0
to 2.5 volts per cell, and this of course means that fewer
cells are connected in series for obtaining a given total
voltage. The capacity of a storage battery is greater for
a given size, and the current is steadier when heavy
draughts are being made on the battery.

The solution is a mixture of sulphuric acid and water, while both the positive and negative electrodes are lead. Zinc and carbon are not used in a storage cell. The plates or electrodes are pasted with a lead salt, and this is different on the positive and negative plates. The plates are generally in the form of a latticework of metallic lead into which the lead paste is forced. When the charging current is passed through the plates and solution, the paste is decomposed and the density of the solution is altered. This difference in chemical composition between the positive and negative plates sets up a difference of potential or voltage of from 2.0 to 2.5 volts, depending upon the state of the charge. When current is drawn from the cell the paste gradually returns to its original composition, and the voltage falls off until at full discharge the voltage is zero. It is best to recharge the battery long before it reaches the fully discharged stage, as the acid is likely to destroy the plate.

Magnetos. A magneto is a sort of alternating current dynamo used for ignition, the magnets being of the permanent type. A coil of wire, known as the "Armature," is rotated between the ends of the horseshoe shaped magnets, and the magnetic flux cuts the wire turns and generates an alternating current in them. The electrical energy in this case is produced by the mechanical energy of the engine, and when the engine stops generation also ceases.

The current thus generated may be of low voltage and passed through an induction coil like the battery current, or the full spark voltage may be generated directly within the winding of the magneto. In the former case the magneto is known as a low-tension magneto, while the latter is a high-tension magneto. The armature winding of a high-tension magneto is very much like that of a secondary coil. It has a primary and secondary winding, and the primary current is interrupted at the instant when the spark is required by an automatic switch known as an interrupter or primary breaker.

In one-cylinder or two-cylinder engines the high-tension current is led directly to the spark plugs from the armature, but when more cylinders are used it generally becomes necessary to use a high-tension distributer, which distributes the current to the various cylinders in the proper firing order. The armature of the high-tension

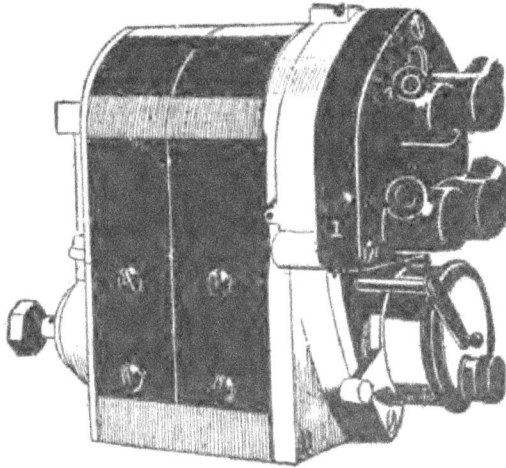

Fig. 13. Typical High Tension Magneto.

magneto must be driven at some fixed relation to the engine speed by means of gears, and the armature must be adjusted so that the interrupter will open just when the piston approaches the end of the compression stroke. This type of magneto is probably the most commonly used types of ignition apparatus, and is not subject to deterioration as with the primary and secondary batteries.

The low-tension magneto may furnish current directly to the igniter in the case of the low-tension make-and-break system, or it may be used with a secondary coil for the high-tension spark. A low-tension magneto is supplied with the Ford car, and supplies current to four high-tension coils mounted on the dash of the car. By means of a two-way switch either the magneto current u the battery current may be used at the will of the driver. The battery current is generally used for starting the engine, while the running is done on the magneto.

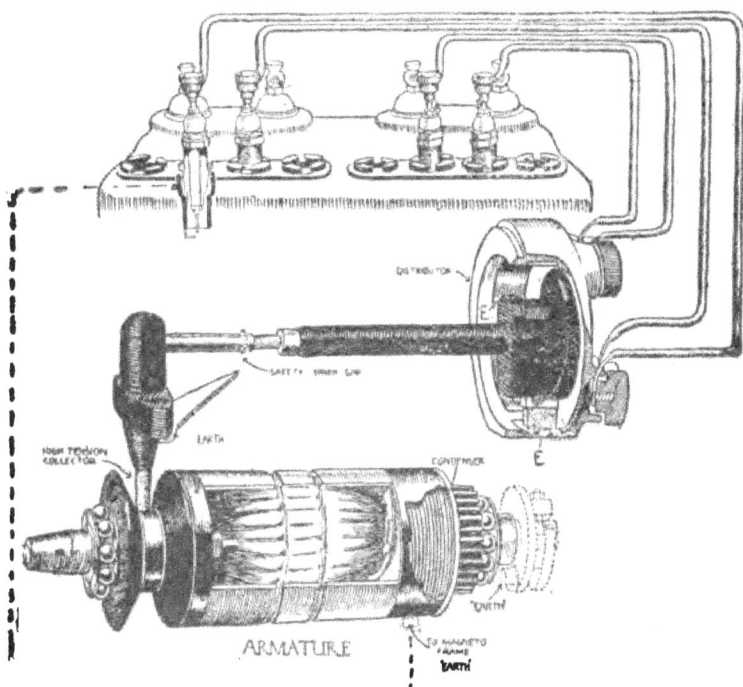

Fig. 14. High Tension Magneto Circuit. The Spark Plugs in the Cylinder Block at the Top Are Fired in Order by the Distributer (E).

Ignition Dynamos. An ignition dynamo is driven from the engine and delivers direct or continuous current instead of the alternating current delivered by the magneto. For this reason it can be used for charging storage batteries, since these batteries can only be charged by direct current. Permanent magnets are not generally used on dynamos, since it is difficult to regulate the voltage for charging, hence the field magnets are electromagnets with coils of wire called the field coils, which magnetize the poles. Since the current in any armature is always alternating, the dynamo must be provided with a device known as a commutator for collecting the armature and rectifying it into direct current for the external circuit. The commutator is a drum built up of insulated copper bars, the bars being connected symmetrically to groups of windings on the armature. Resting on the surface of the

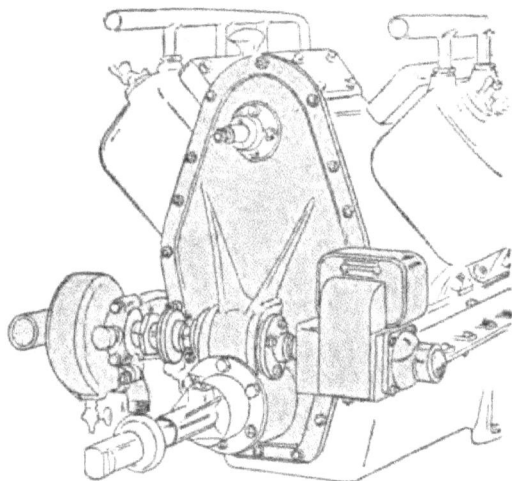

Fig. 15. Ignition Dynamo (at Right) Driven by Gears from Crankshaft of Eight Cylinder Engine.

commutator are bars called "Brushes," which collect the current from the commutator bars in such an order that a positive commutator bar is always connected with a positive brush.

Part of the current is taken from the brushes and is passed through the field coils for magnetizing the stationary field of the dynamo. The current can be used for lighting and for starting the motor, as well as for the ignition, and if a storage battery is used the lights can be used for some time with the engine standing dead. Current may be taken from the dynamo, from the battery, or from both if the dynamo should be charging the battery at that time.

Battery Timers. Battery timers vary greatly in detail, but in general they consist either of a series of stationary contacts, which are successively touched by a revolving arm, or by a cam-operated lever arm, which makes contact with a stationary point intermittently as the cam passes under the cam-arm roller. The latter is the more common on modern battery systems, although the former "Commutator Type" is used on the Ford automobile. The commutator type has as many stationary contact segments as there are cylinders, and a single rotating arm, but the lever breaker is arranged so that there are as many lobes on the cam as cylinders.

Firing Order. The purpose of the multi-cylinder engine is to divide the total power into a number of very small, evenly-spaced impulses so that vibration is reduced and so that there is a uniform flow of power. For this reason care must be taken to divide up the impulses so that they occur at equal intervals, and this means that the primary breaker must measure off equal angles for the contacting arm.

We know that a single-cylinder four-stroke cycle engine gives one power impulses every two revolutions, from which it follows that a two-cylinder engine of the same type will give one power impulse in every revolution, or twice as many impulses in a given time. To find the number of impulses given per revolution by any number of four-stroke cylinders, divide the number of cylinders by two. Thus a twelve-cylinder engine will give $12/2 =$ six impulses per revolution. In the case of a two-stroke engine there are twice as many impulses per cylinder, hence there are as many impulses per revolution as there are cylinders.

Measured on the crankshaft, the angle between the firing impulses will be equal to 360, divided by the impulses per revolution. With a twelve-cylinder engine there will be six impulses per revolution, hence the angle between the impulses will be: $360/6 = 60$ degrees apart. In some engines, notably with the "Vee" type used on aeroplanes and motorcycle motors, the exact theoretical angle may introduce certain practical difficulties, hence the actual angle may be somewhat different than that specified. In most cases it is very necessary to have the engine as

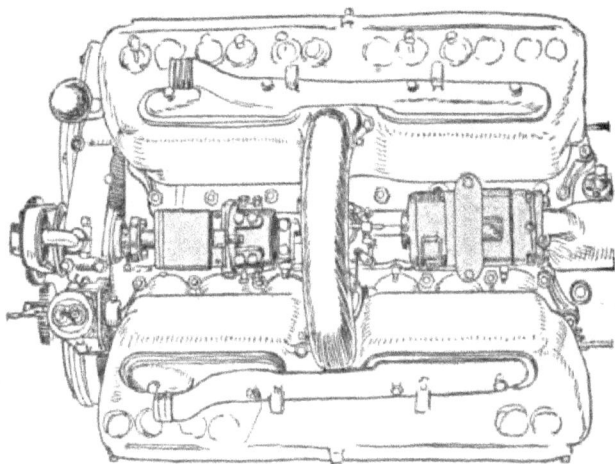

Eight Cylinder "Vee Type" Motor (Plan View), Showing the Two Blocks of Four Cylinders Each, and the Dynamo Between the Blocks at the Right.

narrow as possible, and with an angle of 60 degrees between the cylinder blocks, the over-all width may be too great. The angle of the Liberty motor and the Packard "12" is thus made 45 degrees instead of 60, but the difference in the running and balance is hardly noticeable at high speed.

Vertical and opposed cylinder motors having 2 or 4 cylinders have the crank throws all in one plane with half of the crankpins on either side of the shaft. Vertical engines with 3 and 6 cylinders have the cranks in different planes. That is, a six-cylinder will have the cranks arranged in pairs, each pair being at an angle of 120 degrees with the next pair. If the cylinders are arranged in "Vee" form, a simple straight six-throw crankshaft can be made with the throws all in one plane. A "Vee" type eight-cylinder has the same type of shaft as a four-cylinder vertical, with the cranks all in one plane, but an eight-cylinder vertical requires that the crank throws be arranged so that one-half the throws are at right angles to the remaining half of the throws. The use of more than four cylinders in vertical arrangement requires a complicated crankshaft and the offsetting of the crankpins in the manner mentioned, while the "Vee" arrangement means a short, simple shaft that is comparatively easy to make. All this has an influence on the order in which the different cylinders are fired.

As a general rule, cylinders follow sequence at opposite ends of the engine—that is, the first front cylinder is followed by one as near the rear end as possible, so that the fore and aft rocking moment is reduced to a minimum. Calling the front cylinder of a four-cylinder engine "No. 1," the firing order for the average "Four" will be:

$$1 - 3 - 4 - 2$$

First at the front and then to the back, keeping the moments about the center of gravity opposite and as nearly equal as possible to prevent fore and aft rocking and vibration. If No. 1 was followed immediately by No. 2, both explosions would be on the same side of the center of gravity and there would be an unopposed moment that would cause rocking. By having a front cylinder and then a rear, the moments about the center oppose one another. For perfect opposition No. 4 should follow No. 1, but this is impossible, as the crankpin of No. 4 would be on the wrong side of the crankshaft and the piston would be at the bottom of the stroke instead of the upper end of the compression stroke, as it should be. Cylinder No. 3 is

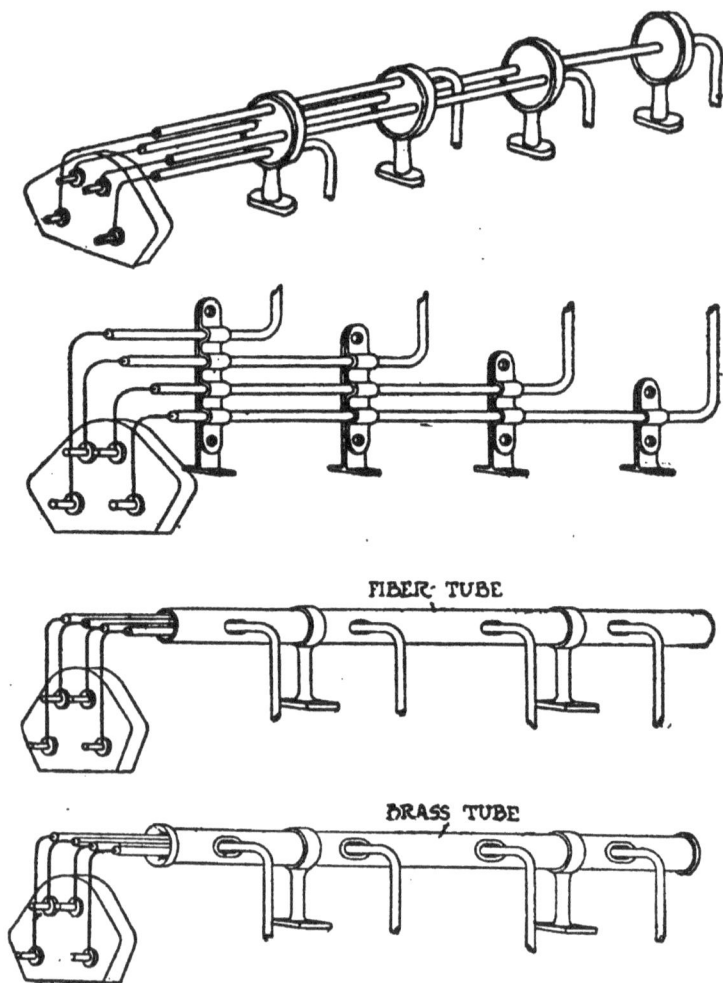

FIBER TUBE

BRASS TUBE

Fig. 12.—Supports for High Tension Wires Leading to Spark Plugs. Disk at Left Represents Distributer of Magneto or Battery Distributer and Wires Shown Leading Out Through Run Are Connections to Spark Plugs. Courtesy of *"The Automobile."*

as near as we can approach the theoretical condition. Cylinder No. 4 follows No. 3, and then follows No. 2 on the opposite side of the center. There are a number of combinations possible.

There is some divergence in the firing order of six- cylinder engines, for the possible combinations are, of course, much greater than with the four. The following are examples of six-cylinder vertical orders:

$$1 - 5 - 4 - 6 - 2 - 3$$
$$1 - 5 - 3 - 6 - 2 - 4$$
$$1 - 4 - 2 - 6 - 3 - 5$$
$$1 - 2 - 3 - 6 - 5 - 4$$

It is very fortunate that the greater percentage of motor manufacturers have established the practice of marking the firing order on the engine. A discussion of several different makes will be found under the chapter on "Timing," page 201.

Advance and Retard. For the best results the combustion should be completed at the end of the compression stroke and before the piston starts out on the downward power stroke. Since the mixture requires some time in which to burn, it is evident that the igniting spark must occur some time before the end of the stroke is reached. When the spark occurs before the center, it is known as an "Advanced Spark." The exact amount of advance depends upon the quality of the mixture and the speed of the engine, for some mixtures burn faster than others, and at high speed the time for combustion is greatly reduced, and hence the spark must be started earlier or must be advanced further than at low speed. Generally, the ignition apparatus is arranged so that the spark position may be varied by hand to meet the different running conditions.

At low speed the engine will pull best when the spark occurs later than at high speed, and in this condition the spark is said to be "Retarded." The time allowed for combustion is now greater than at high speed, and hence the spark is needed only very slightly before the upper center, or in some cases even slightly after the piston starts down on the working stroke. If the engine should be throttled down without retarding the spark, the combustion would be completed long before the piston would reach the top of the compression stroke, and there would be a very strong tendency toward reversing the engine, and hence the power would be much reduced at low speed and the parts of the engine would be severely strained. This condition of an over-advanced spark makes itself known by producing a heavy knocking or pounding in the engine. On the other hand, if the spark is retarded too far the engine will overheat, due to the fact that the flame exists longer during the working stroke, and hence is longer in contact with the cylinder walls. If the spark is kept much retarded, the radiator is likely to boil over and the valves will eventually become burnt or pitted because of the hot gases and flame sweeping over the seats.

Spark Position in Starting. When the engine is being started the spark must be well retarded to avoid "Kickbacks" on the crank. In cranking a motor with an advanced spark, the explosion will take place before the crank reaches the upper dead center, and thus the crankshaft will be turned suddenly and violently in a direction opposite to that of the desired rotation. This is very likely to cause serious injury to the operator, and has been the cause of many broken wrists, loss of teeth and other disagreeable consequences. ALWAYS RETARD THE SPARK BEFORE CRANKING.

Running Position of Spark. When the speed of the engine is increased, the spark should be advanced accordingly, until it is just short of the "knocking point." After the throttle is opened the spark is advanced until ham-

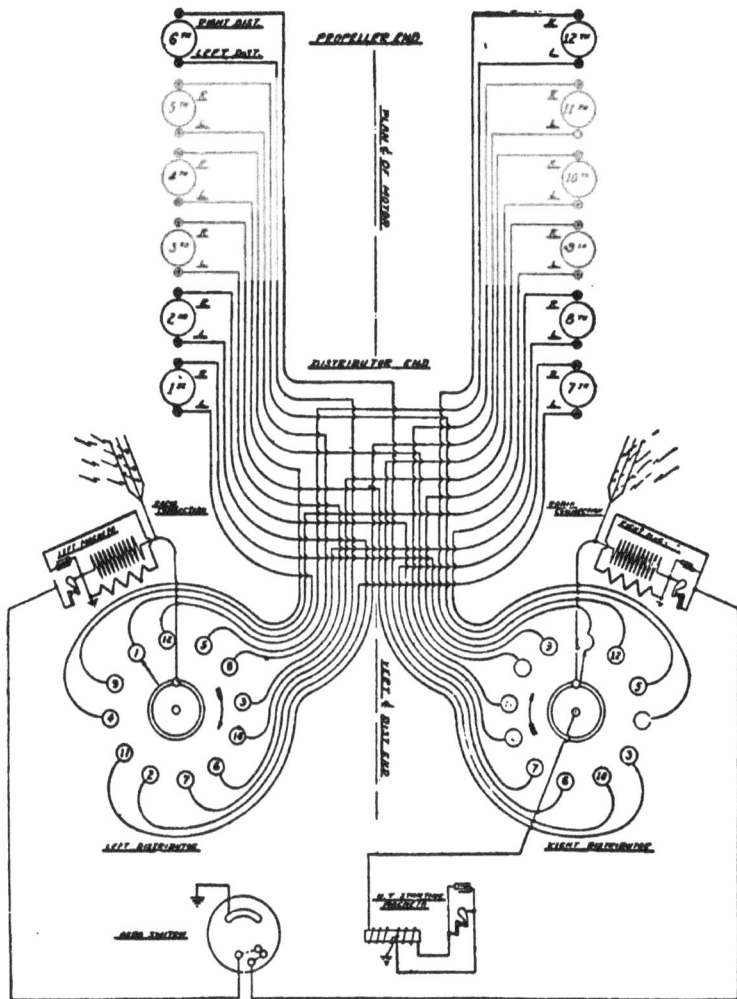

Magneto Circuit Diagram for 12 Cylinder "Vee" Type Engine; Dixie Magnetos.

mering starts, and then is slowly retarded until the knocking just ceases. Whether the spark is first advanced or the throttle is first opened before the spark lever is touched is a matter of opinion, but at any rate the engine should not be run longer than possible before the spark is adjusted to the correct point, or close to the knocking point. Any other position causes a loss in power, heating, or other trouble.

Fixed Ignition. On some automobiles the spark position is fixed and out of the control of the operator. This is to avoid the tendency of the average driver toward keeping the spark in the wrong position, and in the case of a careless operator is often advantageous. The fixed spark is never exactly in the proper position at any speed, but is arranged so that it is a compromise of all speeds, and hence the average results are likely to be better than with a slip-shod control by a careless or ignorant driver.

Ignition Governors. This is a great step in advance over the fixed ignition spark, and at the same time can be arranged so that the control is taken out of the driver's hands. The spark is automatically advanced and retarded in proportion to changes in the engine speed, but, of course, it does not compensate for various qualities of mixtures. The timer or primary breaker is advanced and retarded by some form of centrifugal governor, and advances the spark as the speed rises.

Trembler or Vibrator Coils. In the majority of the older coils, and in a few modern battery ignition systems, a vibrator or breaker is used on the coil, which rapidly opens and closes the primary and thus produces a series or a shower of secondary sparks at the plug. Instead of a single spark at the moment when the timer breaks the circuit, the vibrator causes a series of sparks to continue as long as the timer remains closed. A vibrator coil makes starting easy, and produces better general results with a poor mixture, but is more complicated and difficult to keep in order than the single-spark coil just described. In running some coils are provided with a switch, which cuts out the vibrator when the engine is up to speed and is in normal operation, the vibrator only being used for starting.

The vibrator acts much like the ordinary door buzzer, and is operated through the magnetization of the core. It consists of an iron blade, which is provided with contact points, and is included in the primary circuit in such a way

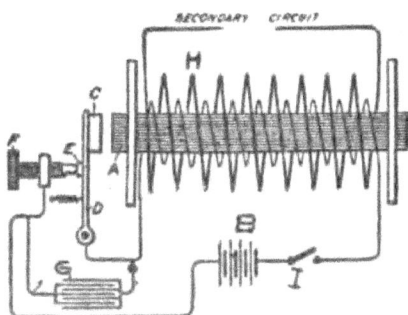

Vibrator Coil in Which H Is the Secondary, C Is the Vibrator Hammer, E Is the Vibrator Contact, B Is the Battery and G Is the Condenser—F Is the Adjusting Screw.

that it opens the circuit when the magnetized core pulls it away from the contact. As soon as the contacts are opened the core becomes demagnetized, and a spring then pulls the vibrator blade back into its original position with the contact points closed. The current now again flows through the primary, the core is again magnetized, and thus pulls the vibrator blade open and again breaks the circuit. This occurs over and over again at the rate of several thousand vibrations per second—in fact, the blade when in proper condition produces a shrill hum.

The contact points on the vibrator blades eventually get hammered out of shape, and the continual sparking finally pits and burns them so that attention must be paid to this point.

Multiple Cylinder Battery Circuit. Fig. 16 shows a complete battery circuit four, a four cylinder motor in which R¹, R². R³, and R⁴ are the spark coils, N is the battery, and Z is the timer or commutator. The spark plugs are shown above the coil and connected to the high tension terminals of the coils S¹, S², S³ and S⁴. The knife switch M cuts out the battery when the engine is to be stopped.

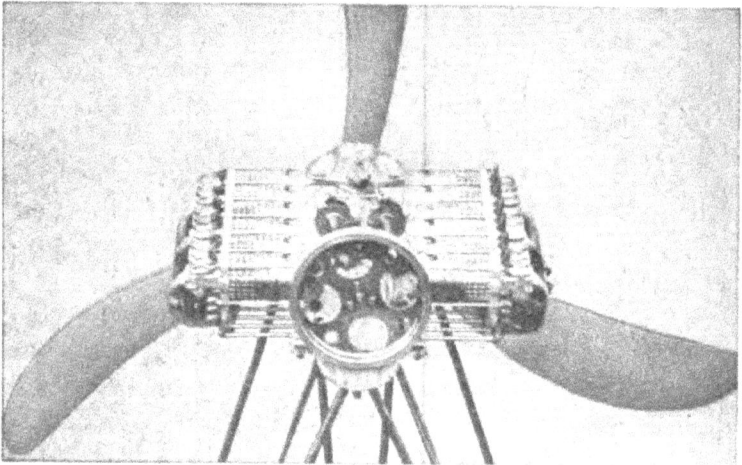

Twelve Cylinder "Opposed" Type Engine, with Cylinders on Opposite Sides of Crankcase.

The timer shaft E carries the arm F, and at X is pivoted the roller contact lever H with the roller at I. A spring G forces the roller into contact with the inner walls of the timer (W-W-W-W) so that the roller makes successive contact with the brass segments (A-B-C-D) while it revolves. Each segment is connected with a coil, and the instant that the roller touches a segment, the corresponding coil sends sparks to a cylinder. The timer housing (Z) can be rocked back or forth by the spark lever (K), so that the spark will occur earlier or later in the piston stroke and thus procure advance and retard

Typical Four Cylinder Battery Coil Circuit Showing the Timer Z, the Jump Spark Coils R1-R2-R3 and R4, the Batteries N, the Switch M and the Spark Plugs V1-V2-V3-V4.

of the spark. When shifted toward the right, the segment meet the roller earlier in the revolution and thus cause the spark to occur sooner (Advanced spark). Turning the housing to the left causes the spark to occur later (Retarded spark). The timer shaft is "Grounded" or connected with the engine frame at (L), and the battery is grounded to the frame at (O) so that the frame acts as a conductor or wire between the timer shaft and the battery and thus saves one wire.

The timer shaft is of course geared to the engine in such a way that the roller moves in exact relation to the piston position.

MASTER VIBRATORS

With multiple cylinder engines trouble is often experienced, when vibrator coils are used, for the reason the spark does not occur in all of the cylinders at the same time. This trouble is particularly noticeable when there is a coil for each cylinder and a vibrator is used with each coil.

Spark variation under these conditions is almost invariably caused by the difference of the weights and tensions in the different vibrators, and by the different electrical characteristics of the coils themselves. When a vibrator or coil does not respond instantly to the contact made by the timer, the coil is said to "lag" behind the current impulse. All coils lag to some extent, as it is impossible to instantly overcome

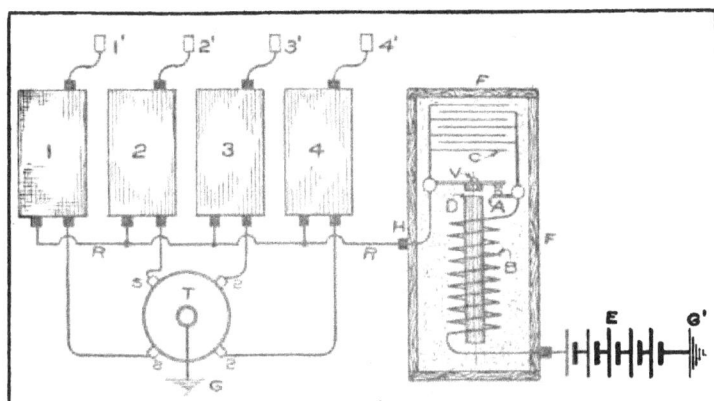

Fig. 15.—Master Vibrator Circuit Diagram.

the inertia of the vibrators or to overcome the self-induction of the primary winding.

With different adjustments, the different coils in a set have different degrees of lag, which results in varying the timing of the spark in the cylinders to which they are connected. With cylinder No. 1, for example, firing regularly on the upper dead center it will often be found that cylinder No. 1 is firing 20 degrees after and that cylinder No. 3 is firing 5 degrees before center. This means that with cylinder No. 1 in the proper position, cylinders 3 and 4 are losing power for the reason that they are not utilizing the compression to the fullest extent. It should not be understood from this that the spark should always occur exactly on the upper dead center for the best results, this point being given simply as an example.

It will be evident from the foregoing that if we used a single vibrator for all of the coils we would have a constant vibrator lag for all, and that the spark would be alike in all of the cylinders, providing of course that the timer was in good condition. This is exactly what is accomplished with the "Master Vibrator" system in which a single vibrator is placed in series with all of the coils through the primary circuit.

The master vibrator generally consists of a small wooden case in which a vibrator, a magnet coil and a condenser are enclosed. The magnet acts on the vibrator in the same way that the primary coil and core acts on the vibrator of a single unit spark coil. The condenser connects across the vibrator terminals in the same way and acts for all four or six coils.

In Fig. 15, 1-2-3-4 are the four coils of a four cylinder motor, and 1^1-2^1-3^1-4^1 are the four spark plugs in the respective cylinders. The timer T is connected to the primary windings of the four coils through the terminals s-s-s-s. The common return R leads from the coils to the master vibrator terminal H, the revolving arm of the timer being grounded at G.

The master vibrator, which is shown greatly enlarged for the sake of detail, is enclosed in the case F. The connection H leads to the vibrator leaf V from which the current passes through the contacts A and the coil B to the battery E. The battery is grounded at G^1. The common condenser C is connected across the vibrator contacts at H and A. The magnet core D is alternately magnetized and demagnetized by coil B as the contacts A close and open. The action of the vibrator, core, condenser, and magnets is the same as in an ordinary spark coil.

It will be seen that the coil B and the vibrator are in series with all of the coils.

Effect of Coil Efficiency on Cells. Inefficient spark coils are proverbally "battery hogs" and should not be tolerated even on the cheapest of automobiles. In the lower grade of spark coils little attention is paid either to the construction of the vibrator or to the proportions of the primary and secondary coils. The following table was compiled from tests made by V. A. Clark, published in the "Gas Review:"

Coil No.	Resistance, Ohms	Current, Amps.	Retail Price, Coil
1	0.619	0.5	$3.00
2	2.732	0.2	1.50
3	0.519	0.6	2.10
4	0.617	0.5	2.75
5	0.562	0.5	1.50
6	2.394	0.2	1.50
7	0.924	0.5	1.50
8	0.814	0.4	1.50

PART II

MODERN BATTERY SYSTEMS

While battery ignition was the earliest used on the automobile it had many objectionable features due to the crude methods adopted in timing the spark and in constructing the coil. Owing to the excessive current demanded by the inefficient coils, to vibrator adjustment, and to the prolonged contact made by the timers, dry batteries would soon become exhausted and would lay down on the job at the most unexpected and inconvenient times. Storage batteries, while giving a more uniform flow of current and requiring less attention, had the disadvantage of being removed from the car during the charging operation. To avoid the trouble of charging the cells at the proper intervals, they were usually neglected until they were well sulphated, which generally meant either a new cell or a thorough overhauling.

With the advent of the electric self-starting and lighting systems all of these conditions have been changed. The lighting battery is always kept in the proper condition by the continual charging of the lighting dynamo so that there is a constant supply of current at a uniform voltage available for ignition purposes. There is no need for removing the storage battery for the reason that all of the necessary elements in its operation are installed in the car. Dry cells are not needed except as an auxiliary or for emergency use, and the troublesome vibrator coil is a thing of the past.

While the magneto proved a great step in advance over the old type of battery system, it too had many failings. It was usually a difficult matter to start a car with an exclusive magneto equipment, it seldom gave a satisfactory spark at low engine speeds, the very time when a hot spark was needed, and when it did get out of order it was almost impossible for the ordinary mechanic to repair.

Under the new conditions the battery system is as capable of a hot spark as the magneto, and the efficiency of the engine is in no way reduced by its use, but as a matter of fact is really increased in the modern types. The battery system will "accelerate" or build up speed more rapidly than a magneto for the reason that the magneto does not give a hot spark at the low starting speed and therefore does not build up as rapidly as the acceleration requires.

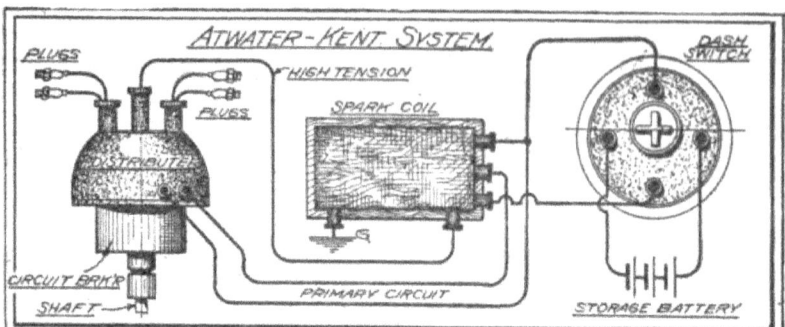

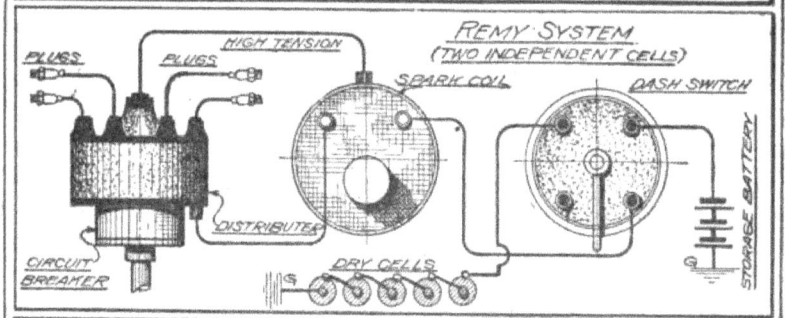

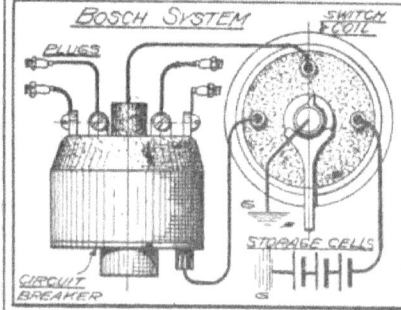

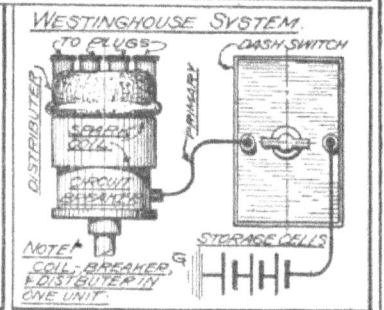

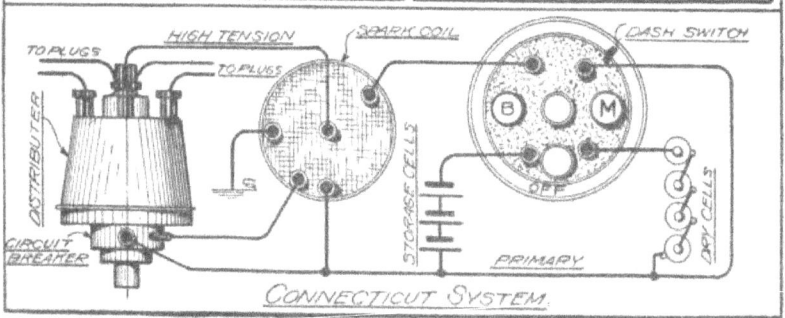

Modern Battery Circuits Commonly Used in Connection
With Self-Starting Systems.

General Description

In general the modern battery system consists of three principal units: (1) the induction coil; (2) the primary circuit breaker; (3) the high tension distributer; these items being, of course, exclusive of the battery.

The principle of the induction or spark coil was explained in a previous chapter, the function of this coil being to increase the low battery voltage to the many thousands of volts required to jump across the spark plug gap.

As before explained, this coil consists of two independent windings, a coil of coarse wire called the "primary," and a coil of many thousands of turns of fine wire called the "secondary."

The primary circuit breaker corresponds to the timer or commutator of the old system, inasmuch as it breaks the primary circuit and causes the spark to take place at the outer end of the piston stroke. The circuit breaker usually runs at cam shaft speed, or what is the same thing, at half crank shaft speed.

In the majority of cases the circuit breaker contact points are mounted in a movable housing so that the points may be rocked back and forth to obtain the retard or advance of the ignition.

With one exception the new battery systems are of the single spark or non-vibrator types, a single break in the primary circuit causing a single spark at the end of the compression stroke. The elimination of the vibrator does away with the mechanical inertia or lag that was so troublesome with early battery systems.

In all types only a single coil is used, whatever the number of cylinders, the high-tension current being led to the different cylinders in the proper order by a revolving switch called the "distributer." This generally consists of a revolving metal blade that turns beneath a series of contacts, there being as many contacts as spark plugs. A lead from the coil carries high-tension current to the revolving blade, which in turn makes successive contact with the contact points in the proper firing order. As both the distributer and the circuit breaker run at the same speed they are generally mounted in one unit for convenience.

In general these systems can be divided into two general classes, those that are of closed circuit type and those of the open circuit type. The closed circuit type have the circuit breaker points normally in contact, completing the primary circuit until the time arrives for the spark. With the open circuit type, the primary circuit is both made and broken instantly at the time of ignition.

With the closed circuit method, the current is on for so long a period that the iron core of the coil becomes thoroughly saturated magnetically, producing a spark of great intensity at the moment the circuit is broken. It is evident that the closed circuit timer is only for use with storage batteries, since the long duration of contact would soon exhaust dry cells. When the circuit is closed and broken almost instantly as in the Atwater-Kent system, the current consumption is very small, which permits the use of the open circuit type with any source of current supply.

Current for the operation of the coil is taken from either the starting and lighting current or from dry cells, or both, if an auxiliary is required to provide for the case in which the storage battery is run down by too frequent starting.

With the closed circuit system in which the battery system is normally flowing provision is made for cutting off the current automatically when the engine stops with the main switch left open through carelessness. This in one case is accomplished by means of a thermostat connected in circuit, the heat produced by a continuous flow warming the thermostat and therefore opening the circuit. In the majority of cases the main switch is arranged so that the current is reversed through the breaker contact points every time the switch is closed. In this way the life of the points is greatly increased, as the metal is not electrolytically transferred from one contact to the other.

THE ATWATER-KENT SYSTEM (Unisparker)

In the first models of the Atwater-Kent its use was confined almost exclusively to dry batteries as its momentary spark was exceedingly economical in the use of current. It consists of one unit comprising the distributer and circuit breaker, and a second unit consisting of the single spark, non-vibrating coil and switch. With this system the intensity of the spark is independent of the motor speed, the duration of current also being independent. Since the breaker is positively driven at cam shaft speed, and as no vibrator is used the spark obtained is practically independent of the battery condition. Normally the contact points are held apart, and at the proper time are forced into contact for an exceedingly brief space of time.

Owing to this short duration of contact, the amount of current taken is exceedingly small, thus making it possible to use a set of dry cells for period several times greater than with the ordinary commutator. As the contacts are normally in such a position that the circuit is open, it is impossible to exhaust the battery by leaving the switch closed. It is impossible for the engine to come to rest with the points of the breaker in contact.

In Fig. 3 is shown the operation of the Atwater-Kent primary circuit breaker, the diagrams being arranged progressively so that the cycle can be more easily seen. A cam S is fastened to the rotating shaft which has a number of teeth cut on its circumference corre-

sponding to the number of cylinders. The trigger T catches intermittently in these notches and is therefore pulled a short distance at
each engagement in the direction shown by the arrow. When the
trigger is released from the tooth it rides up for a short distance

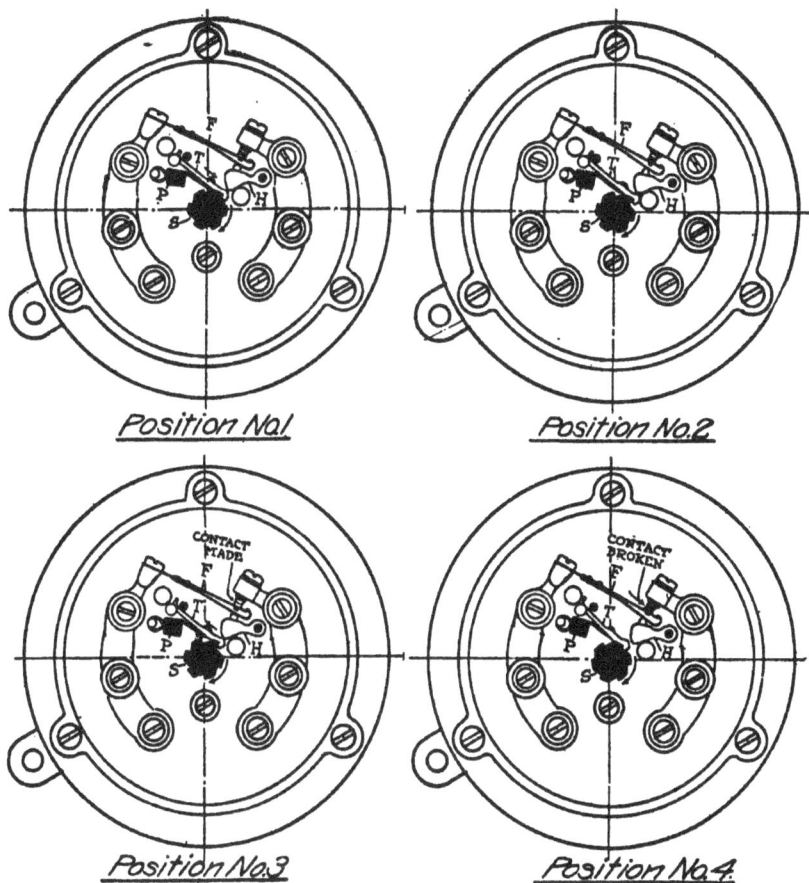

Fig. 3.—Atwater-Kent Primary Circuit Breaker Shown in Four Successive Positions.

on the rounded top of the tooth and then returns to its normal
position through the tension of the spring P.

When the trigger jumps back to its normal position it strikes the
intermediate hammer H which in turn transmits the blow to the contact spring F which closes the circuit. The duration of this contact
is necessarily very short since it is equal only to the time required

for a comparatively stiff spring to snap back into its unstrained position. Tungsten contact points are used for the reason that tungsten has a higher fusing point and is very much harder than platinum, thus giving a minimum of trouble due to pitting or upsetting. In addition to the improvement of the contact points, it should be noted that the current is reversed in direction at every "make" so that the usual trouble due to electrolytic transfer is avoided. Adjustment of the points is affected by the removal of metal shims placed under the head of the contact screw. Since the normal position of T is below and out of contact with the hammer it is impossible for the device to be held in a contacting position. Contact is caused only by the inertia of T forcing it above its normal position. The cam, trigger, and hammer are all of hardened steel so that there is but little wear.

One model known as K-2 is provided with a governor which automatically advances the spark with increasing speeds. This consists of four weights spaced equally around the shaft, the rotation of these weights shifting the cam through an extreme angle of 38 degrees. This angle at crank shaft speed is equal to 76 degrees, measured on the crank circle. The weights are controlled by .springs so that their movement is in direct proportion to the speed of the shaft. In the type provided for manual advance and retard, the entire case is rotated by an external lever.

The high tension distributer is a blade mounted on top of the shaft which receives the high tension current from a central contact point. This blade rotates directly beneath the terminals of the high tension cables. The small gap between the blade and the terminals offers but little opposition to the spark and does eliminate the trouble due to brush wear.

RHOADES BATTERY SYSTEM

The Rhoades battery system consists of a make and break mechanism and a high tension distributor, contained in a single casing, the non-vibrating coil and the attached switch being ordinarily fastened to the dash. The primary circuit breaker operates on the open circuit principle, this being arranged so that if the switch is left closed it is impossible to exhaust the battery. The contacts are always separated when they are at rest. A trigger on the shaft trips over four or six teeth (depending upon the cylinders) and raises a collar, also attached to the shaft. When one of the notches on the collar comes over a projection on the contact springs, the trigger trips over the tooth. This lets the collar down, momentarily forcing the contacts together and then permitting them to spring apart instantly.

In this way, a single spark is produced at each break and as the duration of contact is exceedingly short, there is but little demand on the battery. In fact this is so efficient that weak dry cells may be used with good results. The high tension distributer consists of a radial arm carrying a brass segment which passes in close proximity to the terminals of the high tension leads. There is no rubbing contact at this point and no wear or trouble with deposits of metallic dust that so often cause short circuiting.

THE REMY BATTERY IGNITION SYSTEM

The Remy system is designed for use with storage battery current and can be used either on 6 or 12 volts. It is divided into three units, the circuit breaker, the high tension distributer, and the coil. Referring to Fig. 7, the cam is shown by C, which operates the arm lever A through a hardened follower indicated by F. There are as many faces on the cam C as there are cylinders, each corner of the cam raising the follower lever. The follower lever A carries one contact point which makes contact intermittently with the stationary point held in the block B. The coil spring S normally holds these points in contact. The distributer for the high tension current is mounted above the circuit breaker, and is arranged in such a way that it is stationary at all times and is not affected by the advance and retard of the circuit breaker. This obviates the necessity of moving the high tension wires every time that the advance is altered. A radial arm keyed to the rotating shaft constitutes the moving element of the distributer, the current being fed to the blade by means of a carbon brush mounted in the cover. As with the majority of distributers, the revolving arm does not make actual contact with the high tension wire terminals, but allows the spark to jump through a small air gap between the two sides of the circuit.

THE CONNECTICUT BATTERY IGNITION SYSTEM

The primary circuit breaker of the Connecticut system consists of a pivoted lever actuated by a cam through a roller follower, in a manner similar to a magneto circuit breaker. One contact is mounted in the free end of the lever, while the second contact is mounted in a stationary insulated block. A spring holds these points normally closed. The cam is provided with as many prongs as there are cylinders, each cam prong breaking the primary circuit and causing a spark as it passes under the follower roller on the pivoted arm. Probably the most unique feature of the Connecticut system is the thermostatic switch which breaks the primary circuit should the main switch be left closed while the engine is left standing. A small

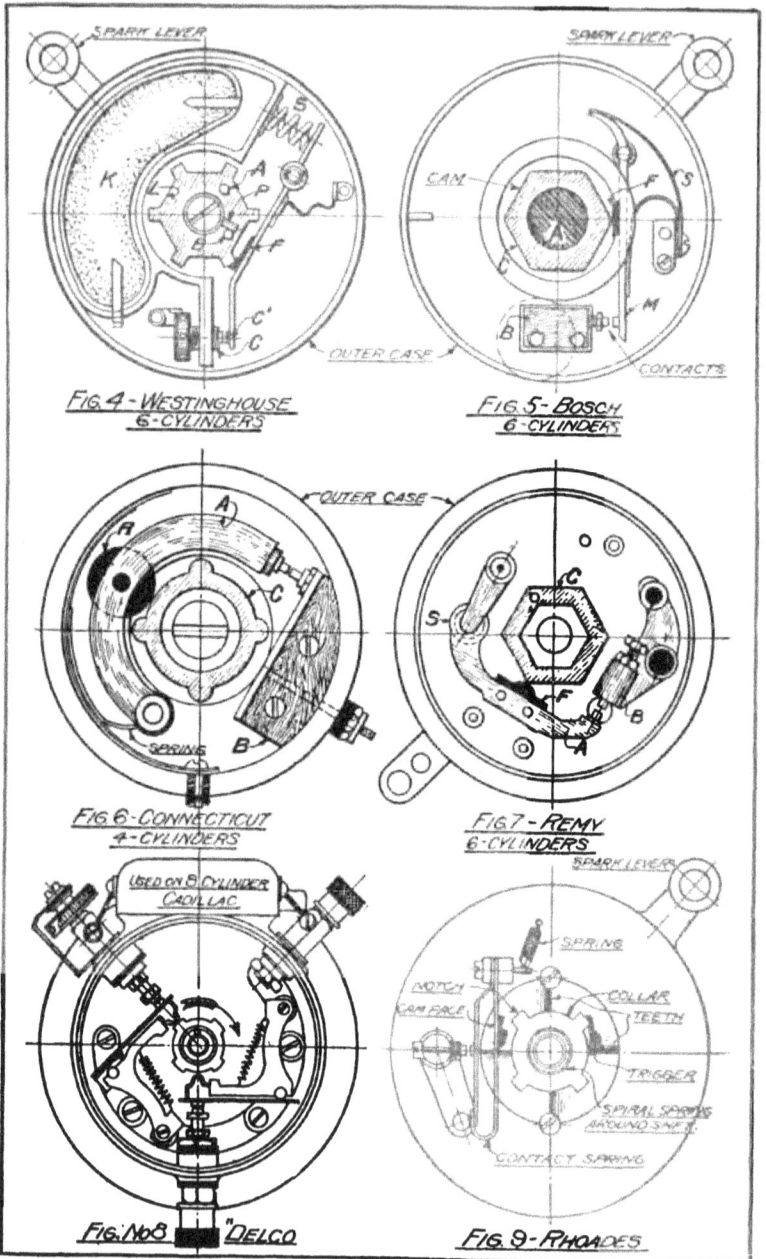

Figs. 4-5-6-7-8-9.—Showing Well Known Primary Circuit Breakers as Used on Modern Automobiles—See Detailed Description.

thermostat is enclosed within the switch proper which heats when the current passes through it from 30 seconds to four minutes without interruption by the circuit breaker.

The heating of the thermostat causes the deflection of a contact bar, which in turn closes the circuit through a small magnet. When the magnet is energized, the armature vibrates the same as the clapper of an electric bell, and coming into contact with the stops attached to the main switch, releases the switch and opens the primary circuit.

BOSCH BATTERY SYSTEM

The Bosch battery system can be used independently or in connection with a magneto. With this system the circuit breaker and high tension distributer are combined in one unit, while the coil is independent. The distributer, which of course runs at cam-shaft speed, can be driven at any convenient point on the engine.

A plan view of the circuit breaker is shown by Fig. 5, in which the driving shaft A carries the cam C, the latter rubbing against the fiber piece F, thus opening and closing the primary circuit. The contacts are mounted so that one is on the movable arm M, while the other is on the stationary block B. The arm M is controlled by the spring S. Advance and retard is obtained in the usual manner.

For starting, the switch is provided with a vibrator, which in turn is controlled by a button in the center of the switch plate. Under normal conditions a quick push on the button will cause a single spark in the cylinder, which is on the power stroke. With a very cold motor a succession of sparks as produced by a vibrator. The vibrator button can be locked in position until the motor is started, after which it is turned to the single spark position.

Four switch positions are provided: B for battery, M for magneto, MB for both together, and O for off.

WESTINGHOUSE TIMER AND DISTRIBUTER

In this system the primary timer, high tension distributer, coil, and condenser are contained in a single unit. It is designed to be operated in a vertical position geared to the cam, or magneto shaft.

The breaker mechanism, which is shown in plan, is quite similar to that used in other Westinghouse instruments except that it is not provided with the automatic advance mechanism. The contacts C and C^1 normally are held together under the action of a small coil spring S; the cam L is mounted loosely on the shaft and is turned by the pin P, which comes against the stops A or B, according to the direction of rotation. The condenser K is mounted adjacent to the breaker mechanism, these two units being beneath the coil and the distributer. The condenser, coil and breaker mechanism are enclosed in a tube of Bakelized Micarta.

The high tension distributer gives a wiping contact. There are two round brushes pressed apart by a small coil spring, one of the brushes making contact with the terminal from the coil and the other rubbing over the six contacts which connect with the wires to the spark plugs. See details of unit on page 44.

TIMING WITH DELCO ON CADILLAC

As supplied to the 1915 Cadillac, the Delco has an ignition relay connected in the dry battery ignition circuit. This breaks the primary circuit immediately after it has been completed by the auxiliary in the timer. This action induces a current in the secondary, giving a spark at the plug. The magnet of the relay attracts an iron bar or "armature," which in turn actuates the contacts which open and close the circuit.

The relay is adjusted by means of a toothed wheel at the top of the magnet, this wheel serving to regulate the distance between the armature bar and the top pole piece of the magnet. Clockwise rotation increases the distance, while counter clockwise rotation decreases it. When the armature is pressed full down the gap is equal approximately to ordinary book paper.

Ordinarily, the adjustment is best performed while the engine is running. Turn the notched wheel in a counter clockwise direction until the motor stops firing, then turn it four or five revolutions in the opposite direction. If the action of the armature bar is feeble when the starting button is pushed, it will be found that there is either dirt between the armature and pole, or that the dry cells are too weak.

As either the battery or magneto systems may require tuning, we will now take up the magneto system. In timing the magneto spark first crank the motor until the piston in No. 1 cylinder is on dead center. Remove the distributer cover, also the rotor, and loosen the adjusting screw just enough to allow the cam to be turned by hand after the rotor is fitted. The adjusting screw should not be loosened enough to allow the cam to turn on the shaft when the motor is cranked by hand. Replace the rotor and turn it by hand until the distributing brush is approximately under the terminal marked No. 1 on the distributer cover.

Switch on the battery ignition, hold in on the vibrator button at the top of the switch on the cowl board and retard the spark lever fully. Pull the spark lever towards advance position and note the point on the sector at which the relay starts to vibrate. If the cam is properly set the relay will start to vibrate just as the spark lever reaches the battery center marked "Bat." C. If the relay starts to vibrate before the spark lever reaches battery center it will be necessary to rotate the cam slightly in a counter clockwise direction to correct the adjustment. If the relay does not start to vibrate until after the battery center on the sector is passed it will be necessary to rotate the cam slightly in a clockwise direction.—"Automobile."

FIXED SPARK

In some motors a compromise between the advance and retard is made, so that the breaker box is held rigid without connection with either manual or automatic control. This does not give the best results, since the spark is overly advanced at low speeds and too far retarded at high because of the varying rapidity of the rate of burning of the mixture at the two extremes of speed. This results in a loss of power, flexibility and fuel where there is much variation in the speed of a motor, as in the case of automobiles. With marine engines, in which the speed is more nearly constant, the bad effects of a fixed spark are not so noticeable.

EFFECT OF PLUG LOCATION

A number of tests in regard to the effect of plug location on the power output have been made recently by C. F. D. Marshall on a six-cylinder motor—

(1) With single plugs over the inlet valves;

(2) With double plugs connected in series, giving two simultaneous sparks over each valve;

(3) With single plugs over the exhaust valves.

The results gave an increase of from 1½ per cent to 2½ per cent over the single plugs with double ignition. With the plugs over the exhaust valves an increase of approximately 1½ per cent was had over the results obtained with the plugs over the inlet. The advantages gained by the plug over the exhaust valve were greater at high speed than at low. This engine was of the "L" type, having the valves all on one side.

In tests on a "T" head motor with the exhaust and inlet valves on opposite sides of the cylinder, D. K. Clark found that the increase due to double ignition was from 10 to 12 per cent.

PART III

MAGNETO PRINCIPLES—MAKE AND BREAK SYSTEM

APPLICATION OF MAKE AND BREAK SYSTEM TO AUTOMOBILES

THE MAGNETO SYSTEM

The next of the divisions into which we separate the systems of electrical ignition is the magneto method, and this can be subdivided into two sections, comprising the low tension and high tension systems. In dealing with ignition by dry batteries and accumulators, it will be remembered that we showed that the spark inside the cylinder can be obtained from a low tension current transformed into a high tension current. It is exactly the same with magneto ignition. We may use either a low or a high tension current, but, as in the case of ignition with dry batteries or accumulators, whenever the high tension ignition is used we must have a low tension current to induce it in an induction coil.

In both systems of magneto ignition the low tension current is generated in the same way, the appliances for this differing principally in constructional details and in the method of wiring up, and the high tension system, as far as magneto methods go, in that provision is made to cause the low tension current generated in the machine to induce a high tension current. For this purpose an induction coil, either separately or incorporated in the machine, becomes necessary.

Low Tension Magneto Ignition.

This being understood, it will be well to deal first with the low tension magneto. That once being thoroughly comprehended, the apparent complication of the high tension system will become quite simple. The principle on which the method is founded lies in the fact that if a coil of insulated wire be revolved between the ends of a magnet, the magnetic influence will so act as to cause a current of electricity to flow through the coil, this being of low tension, and being of a nature suitable for ignition either on the low tension system inside the cylinder or to induce a high tension current in some form of induction coil.

The diagram, Fig. 17, shows the permanent magnets used in a magneto. They are of highly magnetized hardened steel,

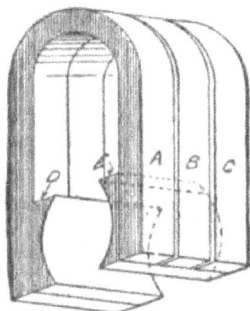

FIG. 17.—THE MAGNETS
OF A MAGNETO
MACHINE.

and are generally arranged in pairs, three sets of pairs being used. These are shown at A, B and C. The efficiency of the machine depends in a large measure on the extent to which the horseshoe magnets A, B and C are magnetized. The arrangement of the magnets in horseshoe form is the most convenient for the purpose of getting the coil, which we wish to revolve, well surrounded by the magnet ends, which are for convenience provided with two soft cast-iron pieces D and E in close metallic contact with the ends of the magnets and

forming a kind of tunnel inside which the coil or armature
will revolve. These two pieces are termed the field pieces
and are often spoken of jointly as the magnetic field.

If a soft iron core of a cross section of the shape shown in
Fig. 18 is taken and a winding of insulated copper wire is

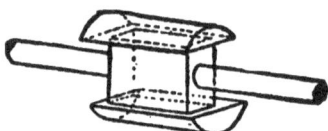

FIG. 18.—THE CORE.

wound around it as shown in Fig. 19, we shall have a coil
which is of a convenient shape to be revolved inside the ends
of the magnets; that is to say, it will occupy a position within
the magnetic field. The object of having this coil revolving
is to cause the magnetic influence, or the lines of magnetic
force, as they are termed, of the magnets to pass through the

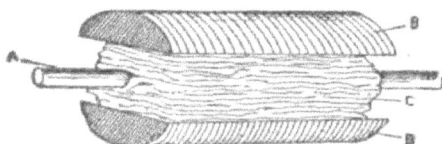

FIG 19.—THE ARMATURE.
A, Armature spindle.
B, B, Iron core of armature.
C, Coil around core.

coil alternately from different directions. This core around
which we wind the insulated copper wire is known as the
armature.

Fig. 20 is a diagram of the ends of the magnets, and a cross
section of the armature in position. AC is the armature core,
and C is the winding of insulated wire around it. F and M
are the field magnets or permanent magnets, and the south
pole of these is at S, the north pole being at N. What effect
the position of the armature with its coil, inside the magnetic
field will have on it, in an electrical sense, we will next en-
deavor to explain as simply as possible.

In describing the soft iron core of an induction coil, we

showed that, when a current of electricity was passed along a winding of insulated wire around the core, the latter became for the time magnetized. A reversal of this idea may be regarded, for the sake of argument, as what takes place in the case of the magneto. Here there is a soft iron core and the winding of insulated wire around it. If, now, we can make this soft iron core a magnet, we can, in a certain manner, induce a current of electricity in the winding around it, but this

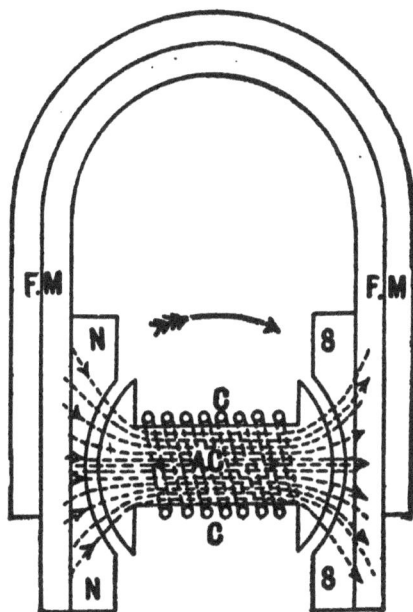

.FIG. 20.—LINES OF FORCE PASSING
THROUGH THE CORE OF THE
ROTATING ARMATURE.

depends on the fact that, to get any results from the induced current in the winding, we must keep reversing the direction of the polarity of the core or armature C. In the case of the soft iron core of an induction coil the polarity, that is to say, whether the north pole shall lie at one end or the other, is determined by the direction in which the wire is wound around the core, and is always the same, there being no practical value or advantage in altering the polarity of the core; but in the present case, in order to get an inductive effect, we may reverse the polarity of the core, or we may cause it to intermittently become magnetized, and this can be done by revolving it between the field pieces. The magnet acts on the soft iron core or armature A C, giving it a magnetic pull which is apparent to the touch, if the armature is allowed to move freely towards the field which is trying to attract it. This it can do when the bearings which hold it centrally are dismantled. This force is represented in our diagram by lines, these lines indicating the direction in which the force acts.

The direction of these lines shows us that the magnetic force is similar to electrical force in that it will always try to act in a direction in which there is the greatest body of metal capable of being magnetized interposed between the

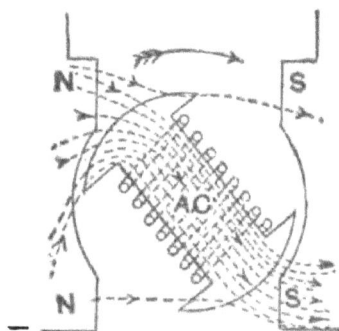

FIG. 21.—LINES OF FORCE IN THE ACT OF BEING CUT.

two field pieces of the magnets. With the core placed in the position shown in Fig. 20 it will be seen that the lines of force are acting as indicated by the dotted lines; that is to say, the magnetic force is acting straight through the center of the armature, and thence through the winding of the coil. Supposing now we rotate the core in the direction of the arrow until it assumes the position shown in Fig. 21, it will be seen that the lines of force have gradually changed their direction. They are still flowing through the greatest body of metal of the core, but they cannot get across in such a direct manner, the direction of the lines of force being indicated by the dotted lines. In fact, there is a kind of leakage, some of the lines of force flowing across between the fields without

4

passing through the armature core, as indicated by the two separate dotted lines.

If the armature is rotated a little further to the position shown in Fig. 22, it will be seen that the lines of force flow across the bulk of metal which forms the sides of the H section of the armature, and do not pass through the center of the armature at all. That is to say, they are not passing through the center of the coil of insulated wire, so that, in fact, stoppage of any magnetic effect on the core has been accomplished by the mere rotation of the core for one-quarter of a revolution in the field. If we rotate the armature another

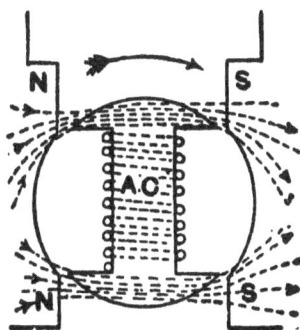

FIG. 22.
LINES OF FORCE PASSING
DIRECTLY THROUGH THE
ENDS OF THE ARMATURE
FROM NORTH TO SOUTH
POLE.

quarter of a revolution in the same direction, it will assume exactly the same position as shown in Fig. 20, but with this important difference, that that part of the armature which was close to the south pole of the magnets now finds itself close to the north pole; so that the direction of the lines of force have been first taken altogether away from the center of the core and then completely reversed relatively to the core itself. It is evident, therefore, that the armature is changing its polarity once every revolution in the field, and it is this constant change of direction of the lines of force through the core that induces the current in the winding around the armature, which current we make use of for ignition purposes.

As the armature A C wound with the copper conductors
or coil rotates between the two poles of the magnet N S it
cuts through the lines of force or magnetism which pass
from one pole to the other, and as a result a current is induced in the coil of the armature. When the armature rotates
through the position as in Fig. 22, it will be seen that the
conductors are cutting the maximum amount of lines of force,
and, therefore, this is the point at which the current is   at
its maximum value.   As it continues to rotate the current
value gradually decreases, and after a quarter revolution
(when it is in the position as shown in Fig. 20) when its conductors are, as it were, slipping along the lines of force,
and not cutting any, its value is nil.   Then it again
gradually grows to a maximum in the reversed direction
(when the armature is again in position as shown at Fig. 22)
and as before, falls to zero, and so on, rising to a maximum
and falling to zero twice in every revolution of the armature.

But something else has also been done by this reversal, for
it will be found that we have changed the direction of the electrical current flowing through the winding; thus we have an
alternating current in the wire.   If we were to connect the
two ends of the wiring C round the armature, and to rotate
the latter at a good speed, we should not get a constant flow
of electricity through the wiring, but an intermittent one, that
is to say, the voltage of the current would start from zero
and gradually rise as the armature revolved and then as rapidly fall.   Two of these, which we may call waves of current,
take place during each revolution of the armature, and if
we break contact in the winding around the armature at the
time when the electrical wave is at the highest voltage, we
shall be able to get a spark.

It is the object of the low tension magneto system to lead
this current generated in the armature winding to the inside
of the cylinder and to break the contact there, and if that
can be done an efficient spark for ignition purposes is produced.

A Simple Form of Magneto—A very simple low tension

magneto system is shown in diagrammatic form in Fig. 23.
One half of the magneto has been cut away in order to show
the armature A lying adjacent to the field B. The coil of
insulated wire around the armature is shown at C, while D is
the armature spindle upon which it revolves in bearings fixed
to either end of the field pieces, the shaft being driven at one
end by a gear wheel which causes it to revolve at some speed
relative to the engine speed. It generally runs at half the
speed, but, in the case of multi-cylindered engines, it becomes
necessary that it should run at such a speed as will give a
diversion of the lines of force at least once every time a cylin-

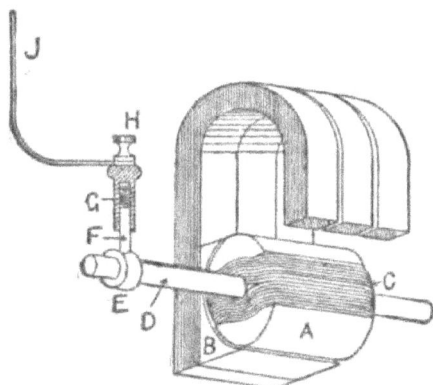

FIG. 24.—DIAGRAM OF A MAGNETO.

der has to be fired. One end of the winding of the coil C is directly and metallically connected to the armature shaft, which,
of course, grounds it. The other end is brought out and attached to an insulated ring E, which is fixed to some part
of the armature or its shaft and revolves with it, but has no
metallic connection with it. On the outside edge of this
ring, and pressing down upon it, is a carbon brush or wiper
F, contained in a tubular case and pressed down by the spring
G. The casing which holds this carbon brush or wiper is
suitably insulated from any metallic part of the machine, and
carries a screw terminal H. It is from this terminal that the
current is taken off by means of the insulated wire J and conveyed to the mechanical contact breaker inside the engine
cylinder.

The Low Tension Igniter—The contact breaker inside the cylinder for use with this type of magneto is arranged in a variety of different ways, according to the ideas of different designers. Its object is first to keep the circuit open, but just before the armature reaches the position shown in Fig. 26 to make contact, and then suddenly break it when the armature is in the position shown. If contact were made all the time there would be no induction set up in the coil, as may be easily understood by the fact that if we ground the winding to the magneto we stop its operation for sparking purposes. Another point is that in the case of multiple cylinder engines it is necessary that only one contact breaker should be making contact at the same time.

In Fig. 24 is shown one arrangement, by means of which contact can be made and broken inside the cylinder. A is

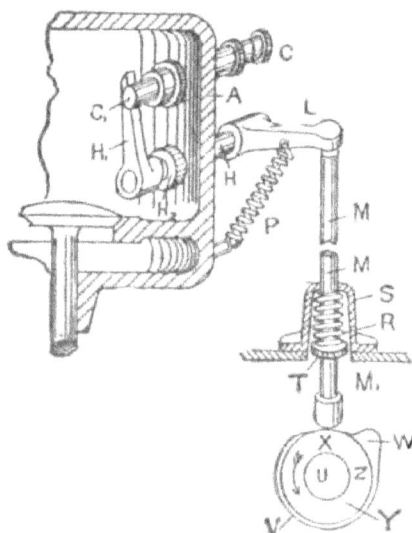

FIG. 24.—LOW TENSION IGNITER.

A, Cylinder.
C, C₁, Igniter plug.
H, Rocking spindle.
H₁, Contact arm.
H₂, Conical joint.
L, Lever on H.
M, M, Tappet rod.
P, Spring.
R, Spring.
S, Thimble over spring.
T, Collar.
U, Cam shaft.
V, Concentric part of cam.
W, Cam projection,
X, Lowest contour of cam.
Y, Cam.
Z, Cam face.

the cylinder in section, so as to show that part of the contact breaker which is inside it, as well as that part which is outside it. Through the wall of the cylinder A passes an insulated plug C, having at its outside end a terminal and terminating inside of a cylindrical piece C1 of steel. This cylindrical piece is insulated from the cylinder by means of the insulation of the plug. The low tension wire from the magneto leads the low tension current to the plug C. H is a rocking spindle, which also passes through the walls of the cylinder. Inside at H_2 it is provided with a tapered ground joint, so that the pressure inside the cylinder will always keep it gas-tight, yet at the same time it can rock in the joint. On its end, inside the cylinder, it carries a lever H1, known as the contact arm. This lever, as will be seen, is so arranged that it can come up in contact with the insulated plug C1. At its end, outside the cylinder, the rocking spindle H carries a second lever L at right angles to the lever H1. A spring P is interposed between this lever and a pin on the cylinder, and this spring keeps the lever L pressed down, and therefore tends to force the lever H1 in contact with the insulated plug C1. This is the normal position of this lever, except under conditions which we shall next describe.

On the engine camshaft U, which is shown in end view, is mounted an ignition cam Y. The shape of this cam is of importance to the operation of the appliance. There is a tappet rod M which passes through the crank case, and at its end carries a hardened shoe M1. It also has a collar T, and is inclosed partially in a thimble S, which is screwed down on to the crank case. Between the collar T and the top of the thimble S is interposed a compression spring R. Obviously this spring will keep the shoe M1 down into contact with the outside of the cam Y. The top of the tappet rod M is arranged to come just under the end of the lever L. The configuration of the cam Y is such as to give this tappet rod a variable movement up and down. In the position shown the shoe is on that portion of the cam which is concentric with the circle Z, which may be described as the lowest part of the cam. In this position it will be seen that the top of the tappet rod M is not in contact with the end of the lever L, so that the spring P can draw the contact arm H1 into close contact with the plug C1. It is just at this position that contact is made, and is about to be broken. The cam Y rotates in the direction indicated by the arrow, so that, after it has rotated a little farther than the position shown, the shoe M1 is suddenly forced up by the projection W on the cam. As it rises it will also knock the end of the lever L up, and therefore will move H1 out of contact with C1, thus breaking contact and firing the charge in the cylinder. The cam continues to rotate, and the shoe gradually drops again as it comes back upon the lower part of the projection W until it reaches that part of the cam V, which is concentric with the shaft, but of large enough diameter to keep the shoe M1 lifted high enough to keep the lever L lifted up and H1 out of contact with C1. During the rotation of the cam the contact breaker is held in this position until it reaches that part of the cam again when the tappet rod M drops, and allows contact to be made, only to be again broken by the projection W on the cam.

The Fiat Igniter—The illustration (Fig. 24) is purely diagrammatic. A drawing of an actual low tension igniter on practically the same lines is shown at Fig. 25, and the same index letters have been applied to this as in the case of the diagrammatic view, so that the reader can refer from one illustration to the other in order to grasp the details of the arrangement. In Fig. 25, the cam operating the rod M is not shown. It is, however, in construction substantially as illustrated in Fig. 24.

The low tension current is led by the insulated wire F to an igniter or plug G1, made of steel, and passing through the walls of the cylinder. It is insulated from the latter by the two cone insulating collars C and C. By means of the nut E, and the two washers D and D, these two coned insulators can be drawn towards each other, forming a gas-tight joint.

The plug G1 is thus entirely insulated from the cylinder wall, but it is in metallic connection with the insulated coil around the armature of the magneto, through the medium of the insulated wire F. It is necessary in order to complete the circuit for the low tension current that some means should be provided for putting the igniter plug G1 into contact with ground, and at the same time providing a means

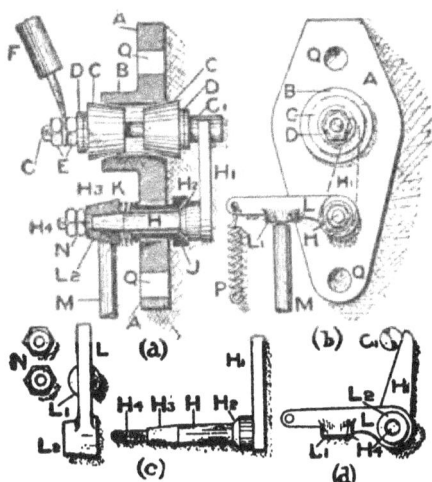

FIG. 25.—THE F.I.A.T. LOW TENSION IGNITION. CONTACT
BREAKER.

A, A, Plate upon which the ignition device is mounted.

B, Boss bored to take igniter, G, G1.

C, C, Composition insulating blocks, tapered to form air tight joints.

D, D, Asbestos and brass washers.

E, Nuts for pulling up insulating blocks C, C, and also for attaching the
high tension cable F to the igniter G, G1.

F, Wire from magneto.

G, G1, Igniter, the end G1 being larger in diameter than the end G, thus
forming a shoulder for the nuts E to pull the insulating blocks up
to cover plate.

H, Rocking shaft.

H1, Contact arm forming portion of H. When contact is broken between
H1 and the end G1 of the igniter, a spark occurs.

H2, Conical facing, forming an air tight joint on the seating in bearing
piece J.

H3, Tapered portion of H.

H4, Screwed portion of H.

J, Bearing piece, in which is formed the seating for H2.

K, Light spring, keeping H2 up to its seating in J.

L, Actuating lever, by means of which contact between H1 and G1 is
broken.

L1, Small boss or facing, against which the tappet or lifting rod M acts.

L2, Boss of actuating lever L.

M, Tappet rod.

N, Nuts, screwed on to the end H4 of H, holding the boss of lever L in
position on H by binding it on to conical portion H3 and by means
of which the adjustment of the angle between L and H can be
made.

P, Spring attached to the actuating lever L, tending to keep H1 and G1 in
close contact.

Q, Bolt holes, for attaching the ignition device to cylinder.

for breaking this contact at the moment when firing in the cylinder is to take place.

This is done by means of the arm H1. This arm is fixed at the end of an oscillating spindle H, which passes through a bearing J in the walls of the cylinder. The outer end of this oscillating spindle is provided with a second lever L, this lever being on the outside of the cylinder. It will be seen that if L is lifted or depressed H1 will also be moved inside the cylinder. The top of the lever H1 is brought into contact with the end of the plug G1, by means of the spring P, at the end of the lever L. This spring holds L down and causes H1 to keep in contact with G1. The low tension current from the magneto can then flow through G1, through the lever H, and to earth through the cylinder walls. M is a vertical tappet rod, its bottom end being in contact with the edge of a cam on the engine camshaft.

What happens so far as the mechanical break goes is this: The cam normally keeps M in contact with the boss L1, on the lever L, and keeps this lever lifted just enough

FIG. 26.

to keep H1 out of contact with G1. Just before it becomes necessary to fire the cylinder, however, the cam allows M to drop out of contact with L1, the spring P draws L1 down and causes H1 to be pressed in firm contact with G1. This takes place just at about the time when the armature is in the position shown in Fig. 21 relatively to the field. As the piston comes to within a very little distance of the top of the stroke and just when the relative position of armature is as shown in Fig. 26 (that is to say, the lines of force have just been diverted, and the edge J of the armature has just left the edge K of the field) the cam suddenly forces the tappet rod M upward; thus M strikes L and lifts it.

PART IV

HIGH-TENSION MAGNETOS

Types of High-Tension Magnetos. To overcome the mechanical complications of the low-tension make-and-break system, the high-tension magneto system has been almost universally adopted on motor cars. Depending on the method by which the low-tension primary current is stepped up into the high-tension current, these magnetos may be classified into three general groups.

(1) Dynamo Type. The dynamo type of magneto may be either of the alternating or direct current type and is generally driven from the motor by a belt or friction pulley in such a way that there is no definite relation between the rotation of the armature and the position of the crank throws.

(2) Transformer Type. The transformer type is geared to the motor so that the armature position has a definite relation to the cranks. A primary circuit breaker is incorporated in the magneto that breaks the primary circuit at the end of the compression stroke. The low-tension primary current generated by the magneto is led to a non-vibrating spark coil. Only a single spark is produced at the time that the circuit breaker opens. This magneto is always of the alternating current type, with two current impulses per revolution.

(3) True High-Tension Type. In this type of magneto the armature generates high-tension current directly without the use of a spark coil. The secondary and primary windings are both on the revolving armature, the high-tension current thus produced being led to the distributer mounted on the magneto.

Direct Current Magnetos. The direct current dynamo is commonly used on stationary engines. As the speed of the device is comparatively high, it is driven with a belt or friction pulley from the flywheel, a governor being sometimes used to keep the voltage at a constant value. It can be used for charging storage cells. A separate circuit breaker or timer must be used since the speed of the armature does not correspond directly with that of the crankshaft. In substituting this type of current generator for a battery of dry cells or storage batteries it is only necessary to disconnect the batteries and reconnect the same two wires with the dynamo.

A vibrator spark coil is generally necessary for each individual cylinder unless a distributer is used. The speed of the dynamo must be very carefully regulated to prevent burning out the coils at a high speed, since this type increases the voltage with every increase in speed. A centrifugal governor mounted on the end of the dynamo shaft which acts by bringing the friction pulley into or out of contact with the flywheel of the motor, depending upon whether the speed is too high or too low.

Alternating Current Dynamos. Alternating current dynamos may be either belt, friction, or gear driven from the motor, and in one or two cases at least are directly connected with the crankshaft. This type is not installed with reference to the crankshaft position and therefore must be provided with a separate timer. No governor is necessary with the alternating current type, as the generator is to some extent self-regulating because of the increasing self-induction at the higher frequencies. This class cannot be used for charging storage batteries. It is placed in the circuit in the same way as the direct current dynamo. Vibrator coils and a timer are used.

In the Ford car a series of magnets placed in the flywheel, and revolving with it, pass a series of stationary coils mounted on a spider. The magnetism threading through the coils, together with the speed of the magnets, generates a low-tension primary current. The magneto is very simple as there are no brushes or contacts, the current being led directly from the stationary coils. Owing to the comparatively high peripheral velocity of the magnets, current is produced at low rotative speeds.

Transformer Type. In this type the primary circuit breaker and secondary distributer are mounted directly in the magneto and in a particular relation to the armature shaft. Only a low-tension current is produced in the magneto, the voltage being stepped up by an independent spark coil. Since the timer is mounted on the armature shaft it is absolutely necessary to time the armature or to gear it to the motor in such a way that the piston and armature have a certain definite relation with one another. It cannot be belt or friction driven.

The high-tension spark coil receives the primary current from the magneto armature and through the circuit breaker in such a way that a single spark is produced at each opening of the breaker. This spark occurs so that the gas in the cylinder is ignited at the end of the compression stroke.

When more than one cylinder is used the magneto is provided with a high-tension distributer, which distributes the current to the cylinders in correct firing order. The distributer is mounted in the upper part of the magneto and is geared to the armature shaft, so that in a four-cylinder motor the distributer travels at camshaft speed. With four cylinders the armature travels at crankshaft speed, and with six cylinders at one and one-half crankshaft speed, the distributer at all times and cases turning at camshaft speed. A high-tension lead from the spark coil connects with the revolving arm in the distributer. Common examples of the transformer type are the Connecticut, Remy, Splitdorf, and the old type Eiseman. At the present time, however, all of these companies with the exception of one also produce true high-tension type.

True High-Tension Type. This is by far the most common type of high-tension magneto for the reason that it is compact and self-contained and is by far the simplest to wire up. It requires no coil except that used for a battery auxiliary.

In the true high-tension type there are two windings on the armature, a primary and secondary, the secondary like the secondary of a spark coil, being composed of thousands of turns of very fine wire. The primary is of coarse wire and is interrupted by a circuit breaker. A spark is produced at every break in the primary circuit. The inner end of the primary is grounded to the frame of the magneto through the armature, while the remaining end of the primary is connected to the inner end of the secondary, the connection to the circuit breaker also being made at this point.

The outer end of the secondary wire is connected to the high-tension distributer through a slip ring mounted on the armature shaft. The distributer is driven from the armature shaft by a gear so that it revolves at camshaft speed. This type is geared to the motor in a definite relation as in the case of the transformer type, the armature shaft running at exactly crankshaft speed in the two and four cylinder types, and one and one-half crankshaft speed in the case of the six-cylinder motor. The primary circuit breaker is then so placed that it opens when the piston is very near to the end of the compression stroke, thus igniting the charge on the upper dead center.

Since two sparks are produced, per revolution of the armature shaft, no distributer is needed with the two-cylinder motor, the connections in this case being led directly from the high-tension slip ring. A lead from each spark plug is brought to the distributer so that as the distributer arm revolves it comes into contact with the terminal of each plug in the correct firing order. A low-tension lead runs from the breaker box to the cutout switch on the dash, so that when the switch is closed the primary winding of the armature is short-circuited, thus stopping the generation of current.

Advance and retard in this type of magneto is had by shifting the casing of the circuit breaker back and forth so that the primary current is interrupted earlier or later in the revolution. In some types the advance and retard is performed automatically by means of a centrifugal governor.

As in the battery coil, a condenser is connected across the terminals of the circuit breaker so as to increase the rapidity of the break in the primary circuit.

TYPICAL TRUE HIGH-TENSION TYPE

In Fig. 1 is shown a perspective view of a typical true high-tension type magneto, the magnets and pole pieces being omitted for the sake of simplifying the drawing. The armature lies between the pole pieces and magnets in the same manner as in the elementary magneto previously described. At the right of the perspective is a section through the armature showing the actual arrangements of the two windings on the armature, the winding in the perspective being simply diagrammatic. The shuttle armature of "H" form is indicated by H in both views.

This armature is connected to the shafts D and N by two brass end plates similar to F. The body of the armature in general is built up of laminated sheet steel to prevent the generation of useless eddy currents and to increase the strength of the magnetic flux through the armature winding. The primary winding is grounded to the armature core at the point Y, and is then given several turns around the iron core K, the outer end of the winding being connected to the connection bolt 2B at the point M. It should be remembered that the primary winding consists of a few turns of heavy wire.

From the point M, the secondary winding consisting of thousands of turns of very fine wire is started. The inner end of the secondary being connected to M makes the secondary simply a continuation of the primary winding. This is not shown in the perspective as it would greatly complicate the drawing, but the true arrangement can be easily seen from the section at the right in which J is the primary and L is the secondary, an insulating strip G separating the two parts of the circuit. The entire series of winding is insulated from the core by the insulation indicated by the heavy black lines. A band I binds the wire against the centrifugal force that tends to burst the winding when the armature is rotating.

Primary current is carried to the circuit breaker jaw 2A and the switch 2D, through the insulated connection bolt 2B, which is insulated from the shaft N by the black insulation shown. The outer end of the high-tension winding is carried to the high-tension collector ring E by means of the insulated pin 2E. A brush at 2B carries primary current to the grounding switch 2D, which when closed grounds the primary and stops the generation of high-tension current. This switch is generally placed on the dash of the automobile.

A primary circuit breaker jaw 2A, which is connected to the primary winding, and is insulated from the shaft, revolves with the shaft and makes intermittent contact with the jaw X at the point Z. The jaw X is grounded to the shaft and revolves with it so that the two contact points are always opposite to one another. Every time that contact is made between the two jaws at Z, the primary circuit is

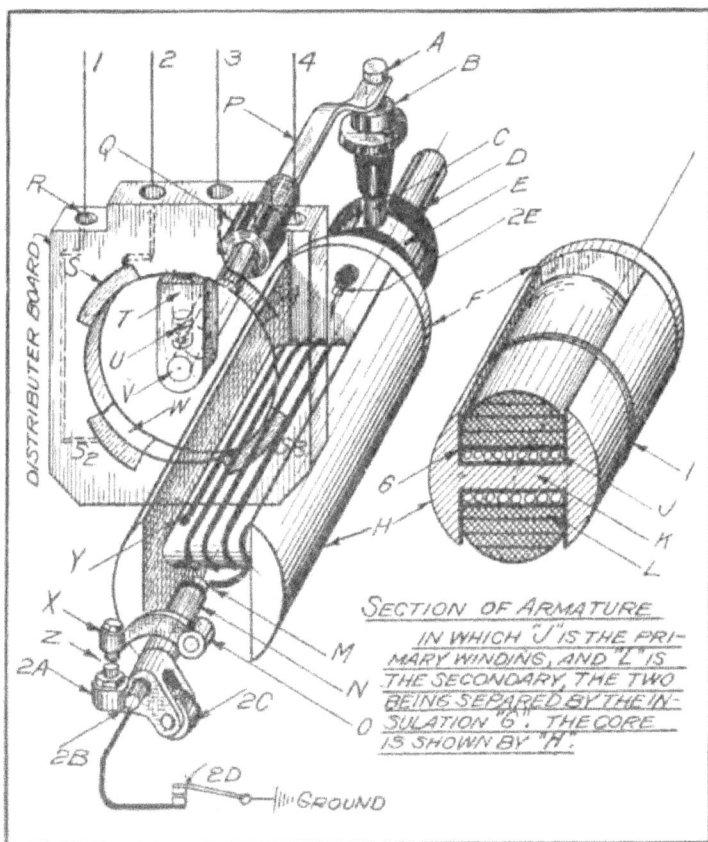

Fig. 1.—Typical True High Tension Type Magneto Showing Construction and Circuit in Diagrammatic Form.

completed through the ground. The opening and closing of the jaws is accomplished by means of a stationary cam which acts on the cam roller 20, the contact between the cam and roller being made twice per revolution.

When the contact is broken, the primary circuit is opened, which gives a heavy current impulse to the secondary winding, this impulse resulting in a spark at the plugs. The spark therefore occurs at the instant when the breaker opens the circuit. The cam that opens the jaws is usually made of fiber board, and is located in the breaker housing that covers the mechanism. In some types of magnetos the cam revolves against stationary breaker jaws, but this is merely a matter of detail and in no way affects the principle of operation. The contact points Z are either of platinum-iridium or of metallic tungsten.

By shifting the breaker housing to the right or left by means of lever, the breaker jaws open sooner or later in the revolution of the armature, causing the advance or retard of the spark. This is similar to the effect produced by rocking the housing of the battery timer. Details of several types of breaker mechanism will be shown in the following chapters.

A distributer board is shown in the perspective which contains the metal sectors S-S2-S3-S4, each of these sectors being connected to the wires 1-2-3-4, which lead to the spark plugs in the cylinders. These sectors receive high-tension current from the brush T contained in the revolving distributer arm V, each sector being charged in turn as the arm revolves. The distributer board is of course built of some high insulating material such as hard rubber or Bakelite, and is shown as if it were transparent so that the armature parts may be clearly seen. A spring U forces the brush into contact with the sectors and also electrically connects the brush with the high-tension current coming through the connector shaft V and the second high-tension brush holder Q.

High-tension current from the secondary winding passes from the connection 2E to the collector ring E, this ring being thoroughly insulated from the frame by the hard rubber bushing D, shown in solid black. The high-tension current is taken from the collector ring by the brush C, through the insulating support B, and to the terminal A. From A the current passes through the bridge P to the distributer arm U through the brush holder Q and the connector V.

The current passes to the plugs through 1-2-3-4, and the plugs being grounded, the current returns through the grounded frame to the armature coil through the arms X and 2A at the moment of contact.

The distributer arm V is driven through a gear (not shown) from a pinion on the armature shaft N. With four-cylinder motors the distributer travels at camshaft speed or at one-half of the armature speed, since the armature of a four-cylinder motor always travels at crankshaft speed.

With a six-cylinder motor, the armature travels at one and one-half times the crankshaft speed, and as the distributer still travels at camshaft speed, the gear ratio between the armature and distributer is 3 to 1. Single-cylinder and two-cylinder magnetos have no distributer, the current being taken directly from the collector ring E. In a type of magneto recently developed for small four-cylinder cycle cars, there is no distributer in the ordinary sense of the word, the distribution being accomplished by two split collector rings. (See Bosch magneto.)

The following table will give the armature speeds for different numbers of cylinders. It should be remembered that in all cases the distributer runs at camshaft speed, and that there are as many distributer sectors as there are cylinders:

(Four-Cycle Type Motors Only)

No. Cylinders	Distributer Gear Ratio	Armature Speed	Note
One	No Dist.	Crankshaft Speed	....
Two	No Dist.	Crankshaft Speed	....
Three	1½ to 1	¾ Crankshaft Speed	....
Four	2 to 1	Crankshaft Speed	....
*Five	No. Dist.	5/4 times Crankshaft Speed	Rotary Motor Dist. on Motor
Six	3 to 1	1½ times Crankshaft Speed	....
*Seven	No. Dist.	1¾ times Crankshaft Speed	Rotary Motor Dist. on Motor
Eight	4 to 1	2 times Crankshaft Speed	Single Magneto
Eight	2 to 1	Crankshaft Speed	Two Magnetos (each 4 cyls.)
*Nine	No. Dist.	9/4 times Crankshaft Speed	Rotary Motor Dist. on Motor
†Ten	5 to 1	2½ times Crankshaft Speed	Radial Aero Type
Twelve	6 to 1	3 times Crankshaft Speed	One Magneto for Twelve Cyls.
Twelve	3 to 1	1½ times Crankshaft Speed	Two Magnetos (each for 6 cyls.)

* Denotes the arrangement used with rotary engines in which no magneto distributer is used, the plugs of the rotating cylinders coming into contact with a stationary brush held by the magneto. The magneto is of the single-cylinder type.

† Denotes a radial arrangement of cylinders, all cylinders being stationary. Seldom used.

TYPICAL TRANSFORMER TYPE MAGNETO

The transformer type of magneto contains a circuit breaker and distributer as an integral part. It must be driven positively at a definite speed, the exact speed in relation to the crank shaft being determined by the number of cylinders in the motor, or the cycle of the motor. A single primary winding Z of heavy insulated wire is placed on the armature, and the inner end is grounded at the point G-3, thus doing away with the necessity of a return wire. The breaker housing L in reality comes directly in front of the armature, but in the drawing it has been placed below so that the armature construction can be more readily seen. The pole shoes P of the magnet embrace the armature in the usual way. A lead from the primary winding connects with a connector bolt G, which passes through the hollow shaft U, the bolt G being insulated from the shaft by the insulating tube. A copper brush E pressed on the end of G by a small spring in the rear, collects the current from the armature and delivers it to the circuit wire terminal 6, from which it flows to the coil terminal T-3. From the terminal T-3 the current passes to the switch contact 1^1, across the switch blade N, where the current splits, part going through the coil and part flowing back to the circuit breaker through 2^1, terminal T-2, and ends at the breaker contact A. A platinum-pointed contact screw M is adjustable in the insulated holder A. It should be noted that the brush E is insulated from the frame by the rubber bushing F.

A rocking breaker arm B swings on the pivot I, to which it is grounded to the frame of the magneto, this arm being swung back and forth by the cam H, which is mounted on the armature shaft U. The cam, rotating periodically, strikes the cam roller K fastened in the arm, opening and closing the contacts mounted in the ends of A and B at the point B. When these contacts are closed the armature circuit is grounded through I to 12, the dotted lines representing the grounded circuit. A pair of auxiliary contacts, C and D, mounted on the back of the rocker and on the timer housing, respectively, are for the purpose of breaking the battery current in the coil.

When the points separate, the current is broken in the primary circuit of the coil, causing a high tension spark. A small helical spring, not shown, pulls the arm B and the roller K, so that it is at all times in contact with the cam H. Since there are two maximum current impulses per revolution of the armature, the cam H is set so that the current is broken twice per revolution at the time when the armature is generating its greatest voltage. The timer housing L may be rocked back and forth by the spark lever 19, by which the spark may be advanced or retarded. Rocking the housing causes the cam H to come into contact, earlier or later, with the roller, thus causing the spark to occur earlier or later in the revolution. The battery breaker C-D is grounded at G-2.

The spark coil, condenser, safety spark gap, the terminals T-1, T-2, T-3, T-4, T-5, and the dash switch are mounted in a wooden box that is usually mounted on the dashboard of the automobile. The battery

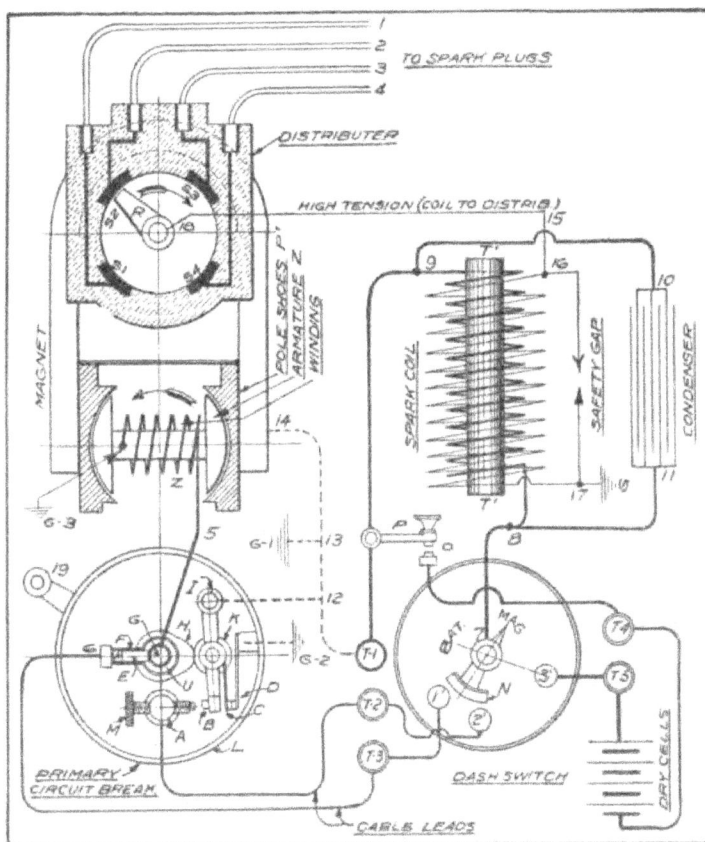

Fig. 2.—Typical Transformer Type Magneto.

is connected with the box by T-4 and T-5, usually marked "Bat." on the instrument. The terminal T-1, marked "3" on the instrument, is grounded to the frame of the machine, while cables from T-2 and T-3, marked "2" and "A," respectively on the instrument, are connected with the stationary breaker contact and with the armature brush E.

Around the soft iron core T-T¹ are wound the primary and secondary windings as shown. In the case of this particular machine, the secondary winding consists of 3900 ohms of No. 34 wire, while the resistance of the primary is only 0.08 ohms, the ratio between the windings being nearly 40,000 to 1. The usual type of tin foil condenser is connected across the primary winding of the wires 9-10 and 8-11, this preventing sparking at the contact points A and B, and acting so as to increase the volume of the secondary spark.

A safety-spark gap is connected across the high tension terminals at 16 and 17, the distance between the discharge points being regulated so that the spark will jump across these points when the voltage becomes excessive at high speeds or in cases when the secondary leads become disconnected from the plugs. Limiting the voltage in this way does away with the danger of puncturing the insulation of the high tension windings. Usually this gap is about ⅛ inch wide, and at speeds above 800 revs. per minute, or with more than 4 cells there is almost a continuous discharge when the plugs are disconnected.

A press button P is used for causing a spark at the plug when the engine is at rest, or for starting on "compression," as it is called. With a warm engine, having its cylinders full of mixture, it is very often possible to start the engine in this way without cranking. The spark occurs when the contacts P and O are separated, the points P and O permitting battery current to flow for an instant through the primary of the coil.

The dash switch is mounted on the front of the coil box and has two switch positions, "Bat." and "Mag." When starting the switch indicator is thrown to "Bat.," and when the engine is firing regularly the switch is thrown to "Mag.," thus cutting the magneto in and the battery out of service. The normal running should always be done on the magneto.

In the sketch the switch is shown on the magneto position, in which the blade N shorts the contacts 1^1 and 2^1, bringing the armature current from 6 to the breaker contact A. At the same time the interrupted armature current is led from the switch at 7 to the primary of the coil at 8, and from the other end of the coil at 9 to the ground at terminal T-1, and thence back to the armature, completing the circuit.

End 17 of the secondary coil is grounded at G, this connection usually being to the lead 9 T-1, as this saves one lead from the box to the frame. The other end of the high tension wire 16 leads through 15 to the axis 18 of the high tension distributer. From the coil box there are the following cables to connect:

2 wires from box to battery (low tension).
2 wires from box to magneto (low tension).

1 wire from box to ground (low and high tension).

1 wire from box to distributer (high tension).

4 Wires from distributer to plugs.

The distributer board, shown in cross-hatch lines, is made of insulating material such as hard rubber or Bakelite. In this material are imbedded four metal sectors, S-1, S-2, S-3 and S-4, spaced at equal distances around the circle. It must be understood that there are as many sectors as cylinders, the present example being for a four-cylinder motor.

High tension current from the secondary of the coil is brought into the shaft of the rotating distributer arc R through the wire 18-15-16. As the arm rotates it comes into contact with the sectors in order and thus connects the high tension current to the spark plugs 1-2-3-4 when contact is made with the segments S-1, S-2, S-3 and S-4, respectively. The distributer thus connects with the plugs in the proper firing order, while the circuit breaker determines the part of the revolution or the time at which the spark is to occur.

The distributer arm R is driven by a gear on the shaft 18 that meshes with a pinion on the armature shaft U, the gear ratio always being such that the distributer arm turns at cam-shaft speed. The gear ratio between the armature and the distributer varies, however, with the number of cylinders used.

The relation of the magneto speed to the speed of the motor or crank-shaft speed depends on the number of cylinders, a single, double and four-cylinder magneto running at exactly crank-shaft speed, while a three-cylinder runs at ¾ crank-shaft speed and a six-cylinder at 1½ crank-shaft. An eight-cylinder will run at twice crank-shaft speed. It must be understood that these speeds apply only to four-stroke cycle motors and to shuttle type armatures which give two sparks per revolution.

Two-stroke cycle motors demand twice the number of sparks per revolution, and for the same number of cylinders as each cylinder in this case fires twice as often. For the speeds of any other number of cylinders see the table under "Typical True High Tension Magnetos." This will also apply to the transformer type.

A type of transformer magneto that was designed by the author is shown by Fig. 3. In this magneto the transformer coil was enclosed in a metal case and placed in the opening between the magnets, thus making the magneto and coil one compact unit and avoiding the use of many wires and cables that are in evidence when the coil is mounted on the dashboard. In the diagram the coil, armature, condenser and circuit breaker are shown approximately in their correct relative positions.

A shuttle armature P is used, one end of the primary winding being grounded to the armature, while the other end is connected with insulated bolt D with the lead K. The heavy line indicates the insulation. A brush B held in an insulating brush-holder A presses on the enlarged head C of the connector bolt D, thus leading the armature current to the external circuit from 11. One lead 10-11 carries the armature current to the primary coil 10-7. Instead of depending on the armature ground connecting with the magneto frame through the shaft and bearings, a separate grounding brush L, held in the metal holder M, was used, this brush grounding the winding at G-2.

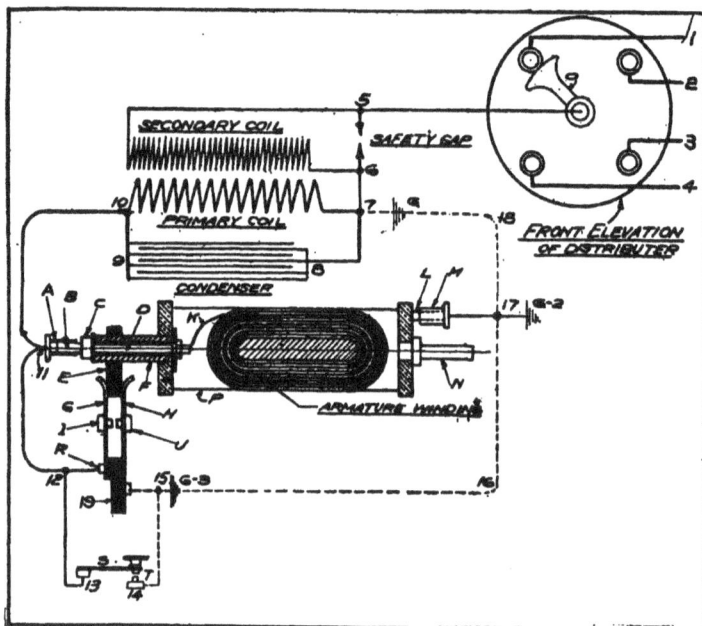

Fig. 3.—Rathbun Transformer Magneto Circuit.

This, as far as the diagram goes, was electrically the same as if the inner end of the winding was connected to the frame, but, mechanically, was much better, as it did not have to depend on a ground through the varying conditions caused by grease or loose bearings. N is the shaft.

At 7 the end of the primary is grounded at G and is connected through the frame at 18, and to brush at 17, all dotted lines representing the grounded circuit. A tinfoil condenser 9-8 was connected across the coil as shown by 9-10 and 7-8. A safety gap 5-6 was connected across the secondary winding, the lead 5-0 going to the high tension distributer arm O. This arm, in rotating, made successive contact with the sectors leading to the spark plugs 1-2-3-4.

The other end of the armature circuit lead from the brush Bat. 11 to the interrupter at R through wire 12. This interrupter consisted of two metal blades G and H, spring tempered, mounted and insulated from each other on the block 19. Two platinum contact points I and J made normal contact with one another, grounding the armature current through 15 at G-3, and from here along the frame 15-16 and 16-17 back to the other end of the armature winding.

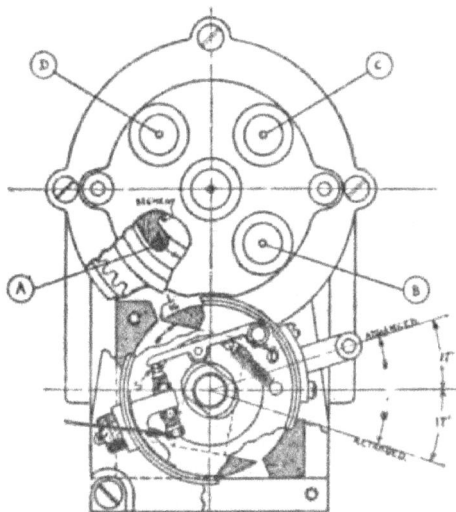

To adjust the points, first clean them with a fine flat file. Then reset adjusting screw "T" so that the gap is not more than .025 inch.

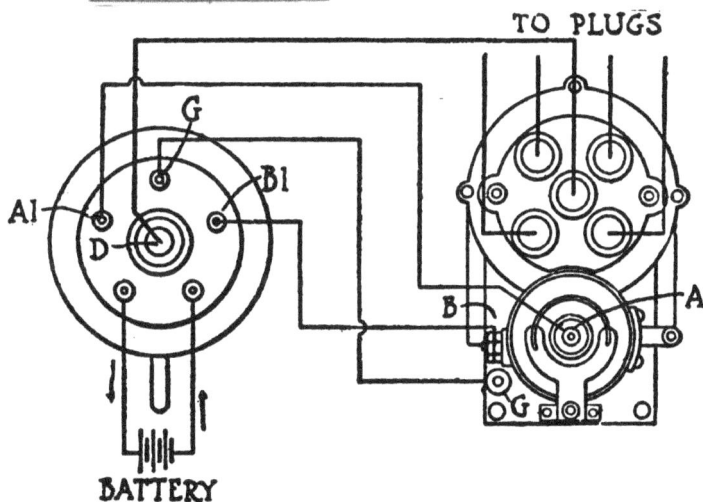

Fig. 4.—Splitdorf Transformer Type Magneto and Circuit. Transformer Coil at the Left of Circuit Diagram. Courtesy of "The Automobile."

A cam E made of insulating material intermittently passed between the contact points at G and H, breaking the primary circuit at I-J twice every revolution and at a time when the voltage of the arma- ture was at a maximum. Every break caused a high tension spark at the plugs. A ground switch S mounted on the dashboard stopped the motor firing by shorting the primary winding across 13-14 and to ground at 6-3.

Leaving the question of the cables to the plugs, the only connec- tion to be made with this magneto was the lead 12-13 to the switch

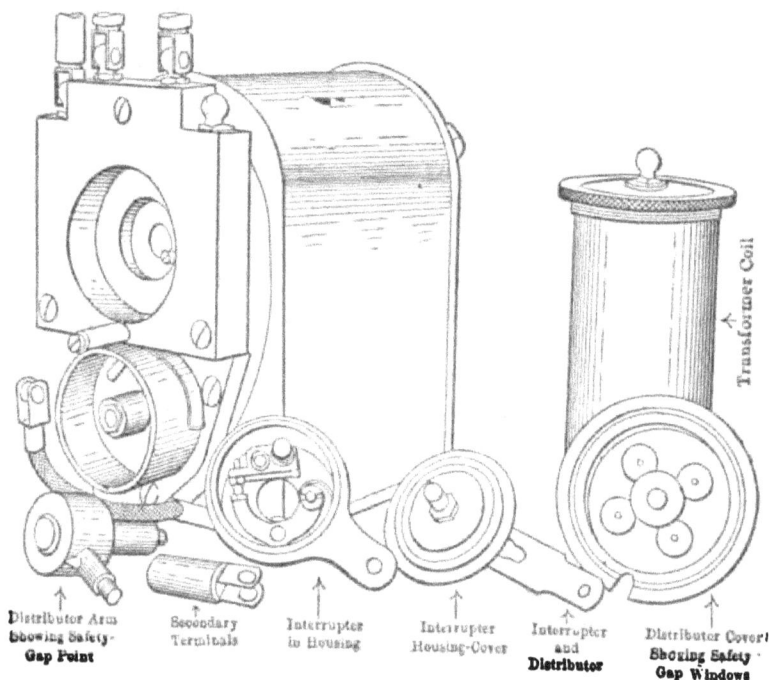

Fig. 5.—Connecticut Transformer Magneto Dissembled.

on the dash, a low tension wire. Very simple when compared to other transformer type magnetos. The number of leads to the plugs would be the same with any magneto.

Another magneto in which the transformer coil is carried between the magnets is shown by Fig. 5. This shows the Connecticut mag- neto in dissembled form, the transformer coil standing at the extreme right of the cut. As in the case of the magneto just described, there are four leads to the plugs and only one lead to the grounding switch on the dash.

The front elevation and circuit diagram of the Splitdorf transformer type magneto is shown by Fig. 4, in which A, B, C, and D are terminals for the spark plug connections. The coil and switch shown at the left in circuit diagram are connected with an auxiliary dry battery which is generally used in starting the motor. A total advance angle of 17 degrees is shown in the elevation with the breaker contacts at the point S.

Mechanical Details.

The Eisemann System—In this system the low tension current is taken away from the magneto and used to induce

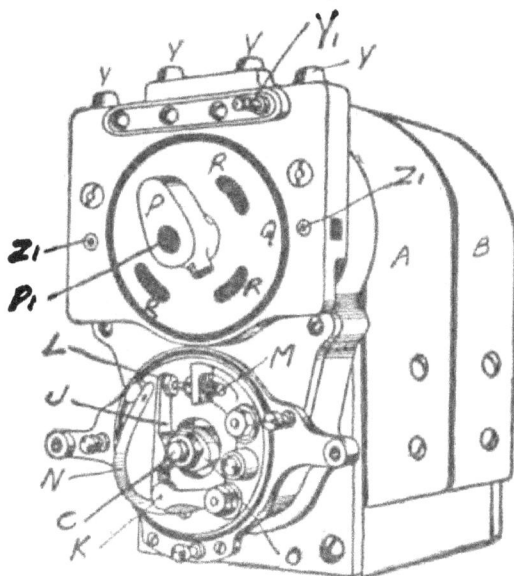

FIG. 69.—FRONT VIEW OF THE EISEMANN MAGNETO.

a high tension current in the secondary winding of an induction coil. This high tension current is then brought back from the coil to be distributed by a distributer forming part of the magneto mechanism. There are several types of this magneto used; they vary generally in their mechanical details, but in principle are all practically alike. In the first place, as regards the armature, this runs on ball bearings in plates affixed to the end of the field pieces, the plates being of some non-magnetic metal. A front view of the machine is shown in diagrammatic form in Fig. 29. The magnets are shown at A and B, and the armature revolves inside between the ends of these, the end of the armature shaft being seen at C. If we look now at Fig. 30 we shall understand how the different units of the magneto are built up. C is the end of the armature shaft. The latter revolves in ball bearings in the end plates and carries the wheel D; this wheel

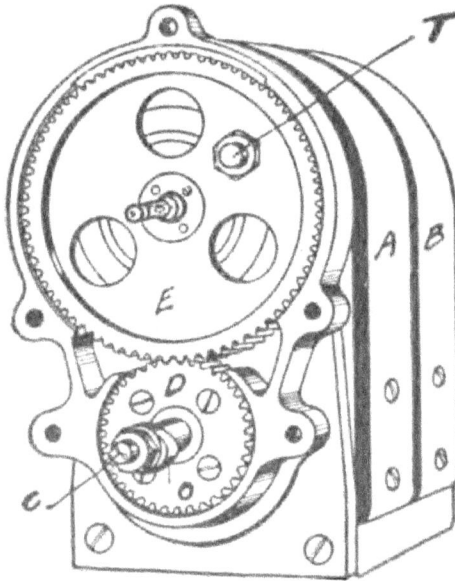

FIG. 30.—THE GEAR WHEELS WHICH
OPERATE THE HIGH TENSION
DISTRIBUTER.

gears with the second wheel E, which is mounted on a second shaft carried in bearings inside the space formed by the horse-shoe magnets A, B. The rotation of the magneto armature will rotate E at half the speed of the shaft. It is the function of E to act as a high tension distributer in a manner which we shall describe later. The contact breaker is mounted on a plate shown in Fig. 29, and consists of a pivoted arm J which is pivoted at K, and is L shaped. At its top end it has a small platinum-pointed head L which comes into contact with the platinum-pointed contact screw M, capable of adjustment, and to which the current is led from the primary winding of the magneto armature. A spring N keeps the pivoted lever J pressed up so that contact is made between L and M. On the end of the armature shaft there is a cam O.. (This cam is seen more clearly in Fig. 30, where it is not confused with the mechanism of the contact breaker.) Now, at the time that the armature is just cutting the lines of force in the magnetic field, this cam comes up against the lever J and knocks it out of contact with the insulated contact screw M, so that the circuit is broken at this moment.

The current from the low tension winding is taken through an insulated contact piece C on the end of the armature shaft, and passes to a carbon spring brush on the cover of the contact breaker, and from there to the platinum screw M of the contact breaker. This carbon brush is also connected to the primary winding of the coil. Thus normally the path of the low tension current is through the contacts of the commutator to ground, through the cam O and the armature, which is connected to ground by a spring-pushed contact piece, not shown in our illustration. This is really to insure a thorough contact to ground. At the moment of breaking the contact the current passes from the carbon brush to the low tension winding of the induction coil and then to ground.

The high tension current is now generated in the separate coil, and is returned by means of the insulated wire to the high tension distributer, also part of the magneto. This is shown very clearly in Fig. 30. At the end of the shaft, on which is mounted the wheel E, is mounted a rotating arm P (Fig. 29). This arm is insulated from the shaft and to it is conveyed the high tension current from the coil. It is the object of this rotating arm to distribute in rotation the high tension current to the different cylinders, which it does in the following way. The current is led to it by means of a terminal Y2 on the cover, which is shown in Fig. 31. This cover encloses the high tension distributer. The two spring fasteners Z, Z engage in the holes Z1, Z1 (Fig. 29). In the center of the cover, Fig. 31, is shown a spring-pushed contact piece G pressed forward by the spring G1. This is insulated from the cover, but is in metallic contact with the terminal inside Y2. When the cover is in position, G presses up against the contact piece P1 (Fig. 29), in the rotating arm, P, and thus the high tension current is led to P, which revolves in front of an insulated disk Q. In this disk are inserted four segments R, R, R, R and a spring-pushed wiper inside the end of the arm P, and pressing against the vulcan-

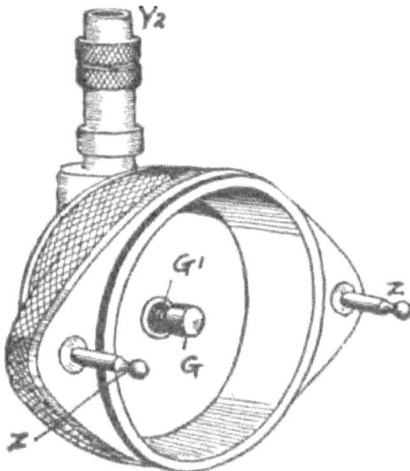

FIG. 31.—THE COVER OF THE HIGH
TENSION DISTRIBUTER.

ite disk Q, makes contact with each of these four insulated segments in turn. These segments are internally connected up to terminals Y, Y, Y, Y on the top of the plate S, and from these terminals wires run to the different sparking plugs. It will be seen that this arrangement will distribute the current alternately to each of four cylinders. In the case of a two-cylinder engine or a six-cylinder engine there would have to be two or six respectively of these insulated segments, and in the case of a six-cylinder engine the ratio of gearing between the magneto and the engine would have to be differently arranged.

TWO POINTS OR TWO-SPARK IGNITION

The greatest power will be developed in a cylinder when ignition and complete combustion occur with the compression pressure at its greatest or when the volume of gas is the least. All events should occur instantly. The time required for the flame to spread through the mixture has made it necessary to have the spark occur before the end of the compression stroke, to reduce the loss of heat to the jacket water and therefore loss of power.

The efficiency and output will increase directly with a reduction in the time required for the combustion, since the longer the burning gas is in contact with the cylinder walls the greater will be the heat loss. Again, delayed combustion is never completed when the compression

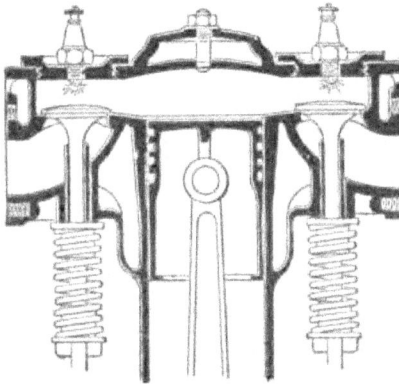

Fig. 6.—Two Point Ignition Showing Two Plugs Over Opposite Valves in "T Head" Motor.

space is the smallest, hence the exposed surface is greater, which again increases the loss.

If the combustion is started at two places simultaneously in such a way that the points of ignition are spaced with equal amounts of mixture or distances between them, it is evident that less time will be required for the flame to sweep throughout the volume. This may be illustrated by the relative times required to burn a candle: (a) lighted at one end, and (b) lighted at both ends.

To obtain the best results, the points of ignition should be well separated, as shown by the accompanying Fig. No. 6. In this case the two plugs are shown over the valves of a "T" head motor located on opposite sides of the cylinder, so that the flame travel is only half instead of the entire distance across the head of the cylinder. This is the reason that "T" head motors give better results with two-point than the "L" head or the overhead valve motor, proper plug location in the two latter types being difficult to obtain.

A better arrangement than that illustrated would be to move both plugs toward the center, so that the distance between the plugs would be twice the distance between the plugs and the wall of the combustion chamber. In this way the spread of the flame ring would only be one-quarter of the distance from wall to wall. This is based on the assumption that the rate of travel from the plugs is equal in all directions. As shown, the plugs are too close to the wall.

A test on a four-cylinder Chester motor, $3\frac{8}{16}$ x $4\frac{1}{4}$ was run at the plant of the Automobile Club of America. The engine was arranged so that single or double sparks could be had by control switch in the magneto circuit. The motor was of the "T" head type.

The greatest power output with single-point ignition with an advance of 4.5 degree was 24 horsepower. With two-point ignition the advance was only 19 degrees for the same output. The maximum power developed with two-point ignition was 28 horsepower with an advance of 32 degrees. This is an increase of 4 horsepower, or 16.6 per cent, over single-point ignition.

Another point noted in the favor of two-point ignition was the fact that a given power was developed at a much lower speed. Should one plug fail, it will be found that the other is generally operative, thus adding to the reliability of the equipment.

EFFECT OF ADVANCE AND RETARD

With all alternating current magnetos there is a definite point in the revolution where the voltage is at a maximum. This point is where the rate of change in the value of the magnetic flux is a maximum, which takes place approximately at the armature position shown by Fig. 7. In this position the magnetic flux is all passing through the armature core and has reached its greatest value in turning through the small angular distance E-F. This angle is very small, as will be seen from the diagram, and if full advantage is to be taken of these conditions, the circuit breaker must open the primary circuit at this point. If the breaker opens at any other point in the revolution, except at the point directly opposite, the spark will be weakened at a given speed.

It should be remembered in this connection that the output, or sparking capacity, of a magneto increases almost directly with the speed, so that a more intense spark is obtained at high motor speeds than at low.

With the magnets stationary, and with the breaker box moved forward or backward from the ideal point in advancing or retarding the spark, it will be evident that the spark is weakened at the extreme points, since the breaker jaws open when the winding is generating a lower voltage. In practice this is exactly what does happen, when the spark is fully retarded, making starting difficult. When cranking the motor a very slow magneto speed is had, which, together with the effect of the retard (always retarded in starting) causes the cranking proposition to be a very difficult one, especially in cold weather when in addition to a poor mixture you have a cold engine and stiff oil.

Even with self-starters, the job with an exclusive magneto equipment is often difficult, for when many starts and stops are made, the

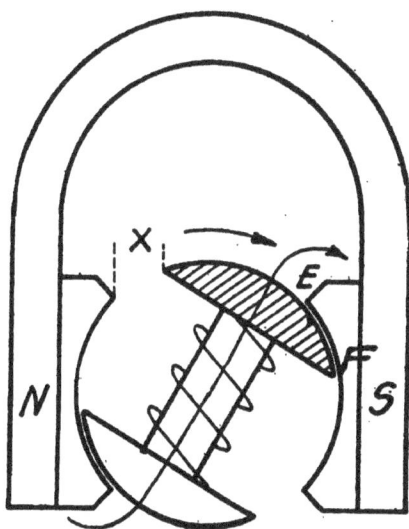

Fig. 7.—Showing Effect of Advance and Retard on Generation of Current.

battery of the self-starter is often nearly exhausted, causing the starting motor to run very slowly. Under such conditions it is usually necessary to fall back on the auxiliary battery system, in which the spark is of the same intensity at all speeds.

During the last few years magnetos have been greatly improved in respect to the intensity at full retard, and many perform wonderfully well when compared with the old models, but even now cranking is still a difficulty in the exclusive magneto system.

With the Eisemann magneto, a special form of tapered pole tips is used, which are efficient at the lowest speeds and greatest retard. In the Mea and "Dixie" magnetos the circuit breakers move with magnets or coils in such a way that the breaker points open in the same relation to the magnetic field at all positions of advance and retard. This, of course, results in an equal spark at all positions of the breaker housing. The K. W. magneto has a spring device which, in combination with a trigger, causes the inductor to "flop over" very suddenly at the sparking point at the slowest possible cranking speeds, thus producing an intense spark in starting. When the magneto is running under normal conditions the automatic spring is cut out of service.

AUTOMATIC ADVANCE AND RETARD

There are magnetos in which the advance and retard is effected automatically by means of a centrifugal governor, the Eisemann Company making a magneto of this type (Type E M A). When this type of magneto is used the advance and retard are out of the hands of the operator, thus giving the correct spark position at all speeds without his attention. The efficiency and performance of the motor is greatly affected in manual control by the ignorance or carelessness of the driver. Overheating and knocking due to an excessively retarded or advanced spark are the common cause of numberless trips to the repair shop.

METHODS OF ADVANCING MAGNETOS

In addition to the method of rocking the circuit breaker housing back and forth to vary the timing of the magneto, there are several other systems that can be and are applied, such as rocking the field magnets or turning the armature in relation to the angular position of the magneto driving shaft.

To understand the working of the two latter types of control it is necessary to bear in mind that the spark is varied in relation to the piston and crankshaft positions. In other words, the opening of the circuit breaker occurs at different parts of the revolution to produce an advanced or retarded spark. Any alteration in the magneto that will cause the spark to occur at different parts of the crankshaft revolution will cause a change in the timing.

Consider a magneto geared to an engine with the magneto mounted in a rocking cradle so arranged that the magnets, frame and circuit breaker can be rocked back and forth as a unit about the center of the armature shaft. If the magnets and frame are now rocked in a direction **against** the rotation of the armature it is evident that the armature will break across the pole shoes sooner than would be the case in its former position, since the pole shoes meet the armature earlier in the revolution. If the circuit breaker is moved directly with the magneto frame, the breaker points will open the primary circuit just that amount earlier, causing an advanced spark. Retard is obtained by movement in the opposite direction.

Since the magnets and breaker move together in this case, the breaker points always open when the armature is at the same position in the magnetic field, that is, the strongest part of the field. By rocking the field and breaker together it is possible to obtain the same intensity of spark at all positions of advance and retard. This method is used by the Mea and Dixie magnetos, as well as by several stationary engine-type magnetos built by the Sumter Mfg. Co.

Stating the above conditions in the form of a rule, "Any relative angular change of the field in regard to a given angular position of the armature will cause a change in the timing." This is true whether the fields are moved in regard to a certain armature position or whether the armature is moved in regard to a certain field position.

With the fields and breaker in a fixed position, turning the armature back and forth on the shaft will cause a change in their relations. This method is adopted in several makes of magnetos in which the armature and driving shaft are free to turn a certain amount in relation. The control of the relative positions is generally had by means of a spirally slotted sleeve which in being moved laterally back and forth on the shaft causes a slight relative angular movement between the armature half and the driving half of the shaft.

This method has all of the disadvantages of the rocking breaker housing type, since with a stationary breaker housing there is relative motion between the breaker and the armature position in the field due to the advance and retard of the cam mounted on the armature shaft. The cam, of course, moves with the armature.

AUTOMATIC SPARK CONTROL

With some magnetos, notably with the Eisemann high-tension type, the advance and retard of the spark is performed automatically. This device when properly set controls the advance to correspond with the varying motor speeds, advancing the spark at high speeds and retarding it at low. Usually the control is effected by the action of some type of centrifugal governor in advancing the breaker housing or in turning the armature in relation to the angular position of the driving or cam-shaft.

The centrifugal governor of the Eisemann magneto is located in a housing at the rear or driving end. Two spring controlled weights actuate a spiral sleeve mounted on the shaft in such a way that the relative positions of the driving shaft coupling and the armature are changed when an increase of speed drives the weights outward by centrifugal force. This is the advance. When the speed is decreased the governor weight springs return the armature to the retard position.

Another type of automatic advance is that of a European manufacturer who advances and retards the armature by a novel form of magneto coupling. The two halves of the coupling are provided with several grooves, a steel ball being placed in each groove in such a manner that the balls normally tend to lie next to the hub of the coupling. The grooves in the magneto half of the coupling are of spiral form, while in the driving half they are straight and radially placed.

When the driving speed is increased, the balls are driven outwardly by the centrifugal force, this movement resulting in a relative movement between the two halves of the coupling, that is to say, the movement of the ball in the straight groove also acts against the sides of the spiral groove so that one half (spiral half) is turned slightly in advance of the other. Since this half drives the magneto armature, it is evident that the armature is advanced with an increase in speed. The circuit breaker housing is stationary at all points in the advance and retard.

There are several important advantages to be gained with an automatic advance, especially in the case of commercial vehicles where carelessness in handling the spark is often a case of efficiency loss and of excessive motor heating and pounding. With a well adjusted device of the class named, the spark position is always at the right point for a certain speed and is entirely out of the driver's control. With pleasure vehicles, the elimination of the spark control adds to the simplicity of the drive and therefore adds greatly to the pleasure of motoring.

In some commercial vehicles an ordinary type of magneto is used in connection with a special centrifugal governor installed by the builder of the motor. The control lever of this governor rocks the timer housing back and forth in a manner similar to that in a manually controlled magneto.

With an automatic timing there is no chance of accident in cranking the motor, since the spark is always fully retarded with the motor at rest. In this way the chances of "kick-back" are greatly reduced.

PART V

SINGLE—DUAL—DUPLEX—TWO-POINT SYSTEMS

MAGNETO WIRING AND CONNECTIONS

When used as an independent source of ignition, the wiring of a magneto is a very simple proposition, but when used in connection with a battery auxiliary, the amateur electrician often becomes confused with the multiplicity of wires and connections. The additional circuits due to a self-starting and lighting system by no means tend to simplify matters.

In general, the circuit of an independent magneto depends upon the type of magneto, i. e., whether it is of the true high tension type or whether used in connection with an external spark coil, since in the latter type there are several primary wires leading from the magneto to the coil on the dash. When this system is "double," that is when two plugs are used per cylinder, the high tension circuit is different than with the single system. To simplify matters, we will confine our attention at present to the combinations commonly used with the true high tension type, or the type in which the magneto windings generate the high tension current without the use of external coils.

Independent Magneto. When a magneto is used without batteries, as in diagram No. 1, there is a high tension lead from each plug P in the cylinders to a corresponding connection post on the distributer D. A primary or low tension wire leads from the circuit breaker C to the switch S located on the dash. The remaining terminal of this switch is "grounded" or connected to the frame of the car or engine. With some late types of magnetos there is no switch S, this short circuiting switch being embodied in the circuit breaker casing, so that the magneto is cut out by moving the spark lever on the wheel to "full retard." When installing the magneto care should be taken to have the magneto base in full metallic contact with the frame of the motor, so that the magneto will also be effectively grounded for the return of the current. The advance and retard of the circuit breaker is shown by A.

Dual System. In the dual system both a battery and magneto are used, the former being used in starting and as an auxiliary against

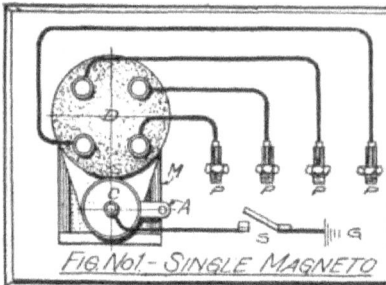

FIG. No. 1 - SINGLE MAGNETO

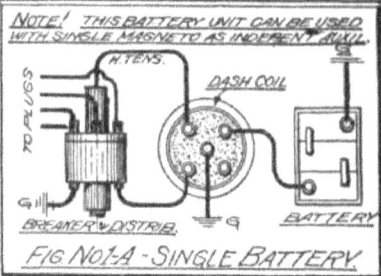

FIG. No. 1-A - SINGLE BATTERY

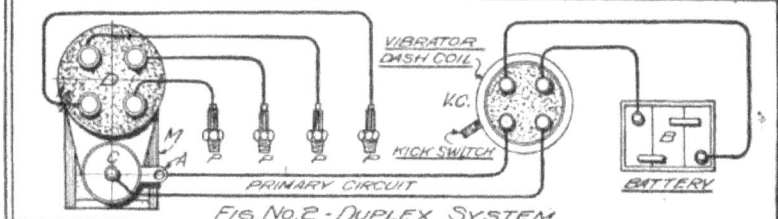

NOTE! THIS BATTERY UNIT CAN BE USED WITH SINGLE MAGNETO AS INDEPEN'T AUXIL.

TO PLUGS · H. TENS. · DASH COIL · BATTERY · BREAKER & DISTRIB.

FIG. No. 2 - DUPLEX SYSTEM

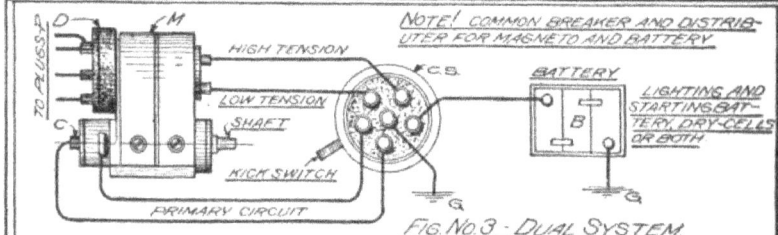

VIBRATOR DASH COIL · K.C. · KICK SWITCH · PRIMARY CIRCUIT · BATTERY

FIG. No. 3 - DUAL SYSTEM

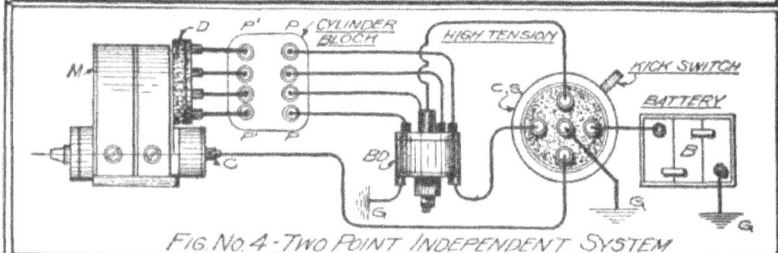

TO PLUGS · HIGH TENSION · LOW TENSION · SHAFT · KICK SWITCH · PRIMARY CIRCUIT · C.S. · BATTERY · LIGHTING AND STARTING BATTERY, DRY-CELLS OR BOTH

NOTE! COMMON BREAKER AND DISTRIBUTER FOR MAGNETO AND BATTERY

FIG. No. 4 - TWO POINT INDEPENDENT SYSTEM

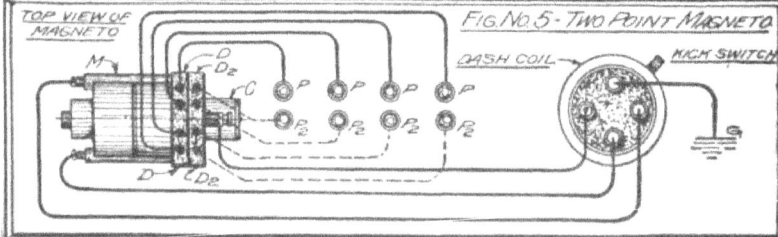

CYLINDER BLOCK · HIGH TENSION · KICK SWITCH · BATTERY

FIG. No. 5 - TWO POINT MAGNETO

TOP VIEW OF MAGNETO · DASH COIL · KICK SWITCH

the failure of the magneto. With the dual system a single set of plugs is used for both the magneto and battery, and the magneto distributer distributes the high tension current for both. The usual connections are shown by Fig. 3, in which CS is the battery spark coil and switch mounted on the dashboard, B is the battery, M is the magneto with the circuit breaker C and the distributer D. And as in the first case, P are the plugs in the cylinders.

As will be seen from the diagram, one pole of the battery is grounded to the frame, as is also one terminal of the coil. For details of this system for different makes of magnetos see the diagrams throughout this chapter.

Duplex System. In the duplex system both the magneto and battery are used, and in some cases an independent vibrator is introduced into the starting system. Instead of having a separate coil for the battery, as in the dual system, the primary and secondary coils on the magneto armature are used to produce the spark when the battery is used. In this type, the circuit breaker and distributer of the magneto are used in common by the battery and magneto. In starting, the switch is thrown so that the battery current passes through the primary winding of the magneto armature, the interrupting and timing being performed by the circuit breaker, each interruption causing a spark at the plugs. The high tension current from the secondary winding of the armature is led to the distributer as in the case when the magneto is working alone.

When running normally on the magneto alone, the battery is cut out of circuit. To increase the spark at starting, the Bosch duplex magneto has a vibrator in series with the armature (see Fig. 2), which is cut out in normal running. In Fig. 2, VC is the combined vibrator and dash switch.

Two-Plug Independent System. To insure complete independence of the battery and magneto systems, the circuits are made entirely separate from one another, as shown by Fig. 4, there being two separate sets of plugs P and P¹, the first for the battery spark and the second for the magneto. Unlike the previous systems, there is a distributer BD and circuit breaker for the battery system, and a circuit breaker C and distributer D for the magneto M. The battery coil CS carries a switch which opens and closes either independent circuit. The battery B is grounded on one side.

This arrangement makes the secondary wiring very complicated and difficult to arrange properly on the motor, since there are twice as many high tension leads to take care of. Since the plugs are the most common source of trouble, the complication due to wiring and the installation of a separate distributer do not make this system advisable in ordinary cases. The comparatively unused battery plugs are generally foul when called upon in an emergency, and therefore the system is little more, if any, reliable than the dual system unless one wishes to assume the trouble of caring for twice the necessary number of plugs.

Two-Point System. To increase the output of a motor, especially on racing cars, it has been common practice to have two sparks occur simultaneously in the same cylinder at rather widely separated points in the combustion chamber. Whether this amounts to any material increase is rather doubtful. A recent test run showed that the increase was only in the nature of 5 per cent, an amount that in

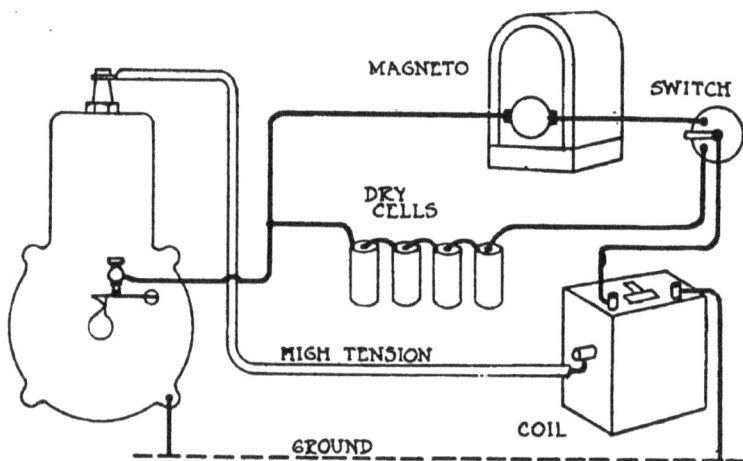

Fig. 6.—Single Cylinder Motor Arranged for Dual Ignition.

an ordinary pleasure car would hardly justify the additional complication and expense.

By installing two points of ignition it was thought that the distance through which the flame had to travel would be reduced, since there were two points from which the flame would spread. An increase in the rate of combustion obtained in this way would naturally decrease the loss of heat to the jacket water, and therefore increase the power. This effect, of course, would be more pronounced in the case of a T-head motor where the distance across the combustion chamber is at a maximum. In the case of the T-head in a certain test this increase amounted to 10 per cent under conditions very favorable to the system, that is, the cylinders were very large, deeply pocketed, and the piston velocity was extremely high. With the automobile in ordinary service the advantages are questionable, especially with L-head or motors having overhead valves.

In general there are two ways of producing the double spark from a single magneto. (1) By providing the magneto with a double distributer, one distributer for each set of plugs and arranged so that each distributer causes simultaneous sparks in each cylinder. (2) By means of a single distributer and special plugs, one plug in each cylinder being of the double pole variety in which both sparking points are insulated from one another and from the metal of the cylinder.

The first method is shown by Fig. 5, in which the double distributers D and D² control the two sets of spark plugs P and P², respectively. The plugs used in this system are of the ordinary type. The primary connections are practically the same as those of the single magneto, and the system can also be used in dual with the battery.

With a single distributer, the high tension circuit must be arranged so that the current passes through the first plug, across the points to the second plug and thence to ground or to the cylinder of the motor. This necessitates, of course, insulating both of the points of the first plug from the cylinder, for if either of the points make contact with the metal, no current will flow to the second plug. The second plug is of the ordinary variety.

PART VI

REPRESENTATIVE MAGNETO TYPES

THE NEW EISEMANN MAGNETO

A new type of Eisemann magneto brought out in 1914, known as model "G," has a number of features not possessed by the model just described. The mechanism of this new magneto is externally completely covered with a sheet metal cover so that it is exceptionally smooth and absolutely oil and waterproof. It is of the straight high-tension type, no coil being used and with no provision for battery connections.

Due to a more efficient arrangement of the armature winding it is possible to use only single magnets instead of the double V's so commonly used in other instruments. The distinctive wedge shape pole piece is still used as shown by Fig. 1, which allows of a hot, strong spark at very low speeds.

It will be seen from the cut that the wedge shaped poles are thicker at the center than at the ends, a construction that causes the magnetic lines of force to be drawn from the ends of the poles to the center. This permits of the entire magnetic volume being forced through the windings with a minimum leakage. In addition, the tapered poles allow of a greater range of advance and retard without a change in the density of the spark.

The high tension slip ring has been moved to the other end of the shaft so that it is now at the same end as the circuit breaker. This change brings all of the delicate parts of the instrument at the same end where they are readily accessible by removing the distributer and circuit breaker covers.

The distributer is entirely new, all of the connections being made inside of the casing, being therefore perfectly waterproof.

By having the distributer disc and high tension collector ring at the same end of the magneto it is possible to have very short high tension connections, the brush from the ring leading directly to the revolving distributer arm.

The high tension leads enter tapered holes in the top of the distributer block, the stripped ends being wrapped around large threaded studs. A keyed washer is placed over the wire for the nut. The forcing of the wires through the holes in the distributer prevents water from leaking past the cables.

The high-tension distributer contacts as well as the ground connection are carbon—light springs being used to insure perfect contact with the distributer arm.

The distributer arm is inserted in the disk with which it rotates touching in turn the high-tension lead contacts. The location of the current collecting slip ring in its present position has made possible the elimination of a large number of parts heretofore necessary. The current collecting brush is mounted in the distributer block and is removed with it, exposing the slip ring when the block is taken off.

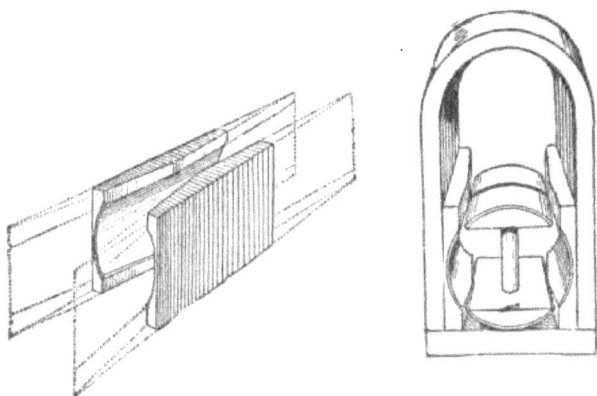

Fig. 1.—Eisemann Poles and Tunnel—"The Automobile."

This makes it unnecessary that the current be led from one end of the magneto to the other as has been the case in the past.

Another excellent feature of the new instrument is that installation and timing have been simplified to the greatest possible degree. On the distributer gear there are two marks, one for left hand engines and one for right hand engines. In timing the magneto it is merely necessary to place one piston in firing position and turn the distributer gear until one of the marks, depending upon the direction of the rotation of the crankshaft, is in line with a screw in the distributer covering.

The circuit breaker, Fig. 2, is entirely new. Instead of being a comparatively heavy arm, there is a very light spring A which carries one of the platinum contacts B and which rotates with the armature shaft; the other contact C is mounted in the part which supports the spring.

The auxiliary spring A1 is merely for the purpose of slightly increasing the pressure between the contact points; it is separated from the main spring by a fibre plug P in order to eliminate the possibility of trouble resulting from static conditions.

Made integral with the breaker box there is a small cylinder D with two fibre inserts E and a third felt insert F, the latter serving merely for lubricating purposes. As the breaker mechanism rotates with the armature, the spring A wipes alternately against the fibre inserts D and E, thus making and breaking the primary circuit. The simplicity of this mechanism is only one of its noteworthy features. As there are no bearings, it is impossible for wear to cause irregular firing, and as there is nothing to stick possibility of trouble on this account is

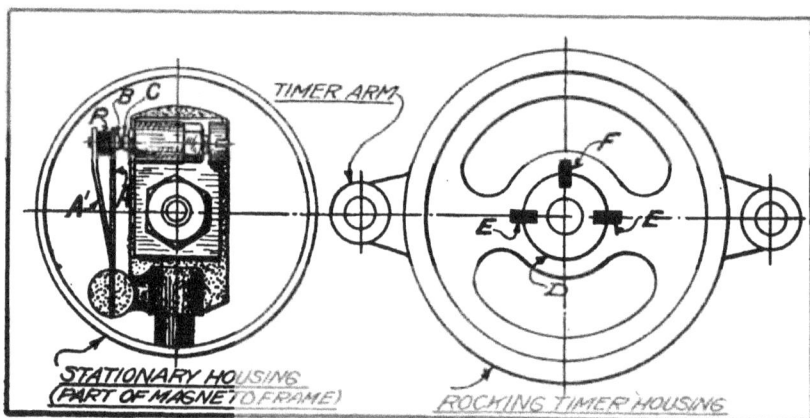

Fig. 2.—Eisemann Circuit Breaker.

positively eliminated. Another very valuable feature is that owing to the exceptional lightness of the parts there is no battering of contacts, which, consequently, may be expected to wear a correspondingly longer time.

BOSCH HIGH TENSION MAGNETO

The Bosch DU4 type is a typical true high tension magneto, the armature containing a primary and secondary winding, the primary being periodically interrupted by means of a circuit breaker. The circuit diagram is shown clearly by Fig. 3, which will serve as a guide to the actual construction that will be described later on. For clearness, the armature is shown in side elevation, while the distributer and circuit breaker are front elevations. The primary wiring is shown by solid heavy lines, the secondary by fine solid lines, and the grounded circuit by dots and dashes.

Since the secondary winding is simply a continuation of the coarse wire primary winding it is shown as a single coil wrapped around the core of the armature. The high tension is collected at the left of the armature by means of a collector ring and brush, the lead from the upper terminal of the brush being connected to the safety spark gap on its way to the distributer brush.

The distributer brush as it revolves makes successive contact with distributer segments 1-2-3-4, leads from these segments, running to the respective spark plugs 1-2-3-4 shown in the upper left-hand corner of the diagram.

A condenser is housed with the armature at the right whose pur-

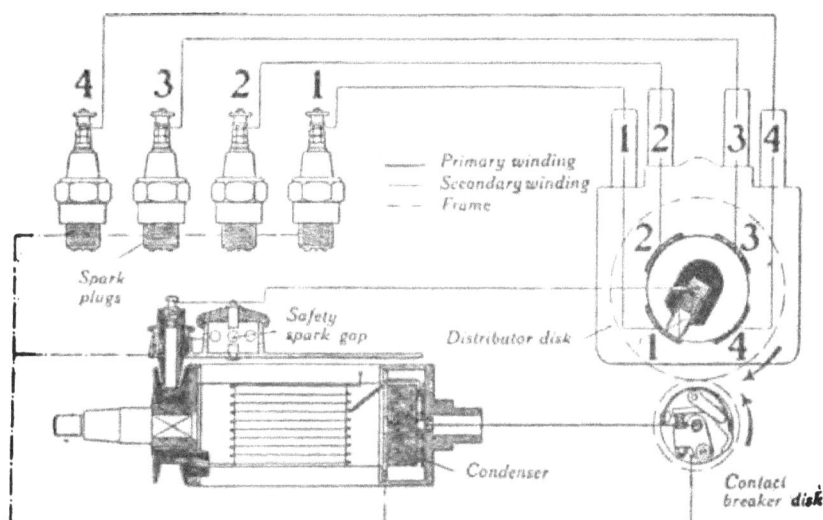

Fig. 3.—Bosch High Tension Magneto Circuit.

pose is to absorb the spark at the circuit breaker points and to hasten the rapidity with which the primary circuit is broken. One end of both the primary winding and the condenser is grounded, the remaining condenser terminal being connected to the lead that runs from the armature to the circuit breaker at the right. The outer shells of the spark plugs, the frame, and the armature are all grounded as will be seen from the dot and dash lines.

In the longitudinal section shown by Fig. 4, the high tension current from the secondary winding is led to the high tension collector ring 9. A brush 10 pressing on this ring collects the current, and through the spring 11, the bridge 12, and the brush 13, it passes to the rotating distributer brush or arm 15. In rotating, the brush makes successive contact with the distributer segments previously mentioned in the circuit diagram. All brushes 10-13 and 15 are pressed against their bearing surfaces by small spiral springs.

A serrated edged terminal shown projecting from the bridge 12 into the safety spark gap housing is placed opposite to another terminal fastened to the top plate of the armature tunnel. This gap prevents an excessive voltage that might be caused by a loose or broken high-tension connection.

Primary current is led from the armature to the circuit breaker through the insulated connection bolt 2, an intermediate connection being made to this bolt from the condenser 8. The outer end of the bolt 2 is connected to the interrupter or circuit breaker jaw 3, an insulating strip 4 separates the block from the metal of the frame. At the end of 2 a spring controlled brush carries current to the terminal 24 through the spring 26 and the clip 25. A connection from 24 is led to the grounding switch on the dashboard whose purpose is to stop the engine by grounding the primary current. The supporting block 27 is insulated from the clamp 23 which holds the distributer cover 22 on the distributer disk 16.

A hard rubber hub 14 carries the brush 15, these parts being readily accessible through the removal of the distributer cover 22. The prolonged shank of the hub 14 rotates in the bearing at the left of the hub, the bearing being thoroughly insulated from the current conducting rod that runs from the brush 13 to the brush 15. Above the hub in the open space will be seen a face view of one of the distributer segments.

Under the base of the magneto and directly under the condenser will be seen the primary grounding-brush through which the current returns to the armature from the grounded parts on the frame. This avoids passing the primary current through the ball bearings, and so saves them from the pitting action that would be certain with the passage of current through the ball races.

The end of the primary winding is connected to the plate 1 into which the connecting bolt 2 is screwed. This plate is of course insulated from the frame by the strip of hard rubber shown between the end piece and the condenser 8.

Current from 2 enters the breaker block or jaw 3, which on referring to the front elevation, Fig. 3, will be seen to carry the platinum breaker point retained by screw 5. These parts are both insulated from the breaker disk 4 which carries the rotating parts. A contact breaker lever 7 (grounded to the frame) carries a platinum screw 29 which makes contact intermittently with the first platinum point 5. These points are normally forced into contact by the flat spring 6. It should be remembered that the contact block 3, points 5 and 29, the lever 7 and the spring 6 are mounted on the armature front plate 4 and revolve with it.

Two fiber cam disks 19-19 mounted in the breaker housing make contact with the toe end of the lever 7, causing the platinum points to open every time that the end of the lever passes the cams. As this is a shuttle armature giving two current impulses per revolution, there are two cams to open the breaker at the highest voltage peak of each impulse. The points are therefore opened twice per revolution, giving a high tension spark at the plugs at each current interruption. A rocker arm 20 is connected with the breaker housing so that the housing and the cams can be rocked to and fro for advance and retard.

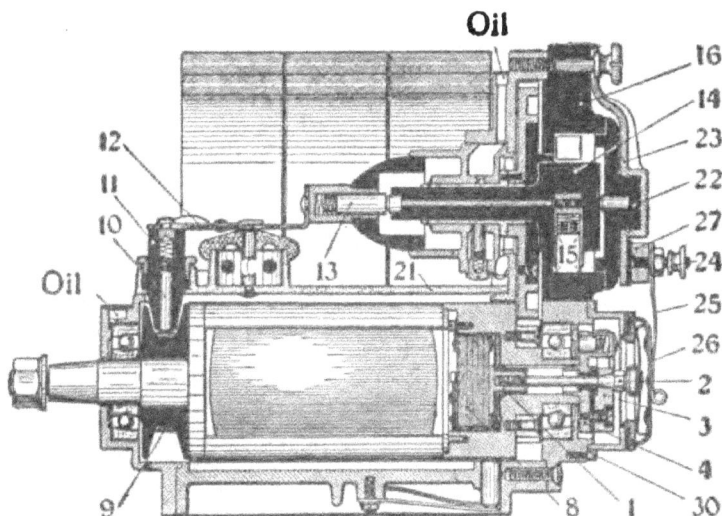

Fig. 4.—Longitudinal Section Through Bosch Four Cylinder High Tension Magneto.

When the house is turned against the rotation of the armature, the toe of the lever 7 meets the cam 19 earlier in the revolution, causing the interruption to occur earlier and consequently advancing the spark.

Since the lever 7 is grounded it is in connection with the beginning of the primary winding which is also ground. As the block 3 is connected to the outer end of the primary, the circuit is closed when the two platinum points are in contact, allowing the current to flow and build up the magnetic flux. When the points separate the flux contracts suddenly through the secondary windings, causing a high potential in the secondary.

The brush 15 carried in the distributer arm 14 receives the high tension current as described in an early part of this section. The distributer segments connect with plug sockets 16 into which are pushed the plugs or spring jacks 18 that carry the high tension cables to the spark plugs in the cylinders.

There are as many segments and plugs as there are cylinders. With single and double cylinder engines there is no distributer, the high tension current being carried directly to the spark plugs from the high tension collector rings. In every other respect the construction is the same.

The distributer brush is driven from the armature shaft by a gear and pinion, the distributer traveling at half the armature speed with a four-cylinder engine and at three times armature speed with a six-cylinder.

With the shuttle type magneto of which the Bosch is an example there are two sparks per revolution of the armature. For the proper ratio of the armature to the crank shaft speed of the engine see the table under the "Elementary True High-Tension Magneto."

As a guide to the proper magneto speed, find the number of impulses given by all of the cylinders in one revolution, Now find the gear ratio that will give the magneto a current impulse for every power impulse of the engine. With a four-cylinder, four-stroke cycle engine there will be two power strokes per revolution, since each cylinder has a power stroke in every two revolutions. Since the magneto also gives two sparks per revolution, it will run at crankshaft speed.

Needless to say, the Bosch magneto must be geared or chain driven by the engine, since there is a positive relation between the piston position of the engine and the time at which the circuit breaker opens the primary circuit.

BOSCH OSCILLATING MAGNETO

When very low rotative speeds are used as with large, heavy duty stationary engines, it is difficult to obtain good results with the ordinary type of magneto. Under these conditions it is usual to use a magneto in which the armature is snapped past the fields at the firing point by means of a spring and tripping device, the armature being oscillated back and forth instead of being rotated continuously in one direction.

An oscillating magneto built by the Bosch Magneto Company is shown in outline by Fig. 5. The armature is actuated by a rotating cam which moves the armature 30 degrees from its normal position when at rest. When the face of the cam "b" strikes the trip lever, the armature is turned until finally it assumes the position shown; a very little farther motion will release the lever and allow the springs

to snap the armature back in a left-handed direction. The direction of
cam rotation is shown by the arrow.

It will be seen that the armature speed and hence the intensity of
the spark is absolutely independent of the engine speed.

A double wound shuttle armature is used as with the usual true
high tension type, and the primary current is interrupted by the usual
form of circuit breaker. At the moment of interruption the spark
appears at the plug.

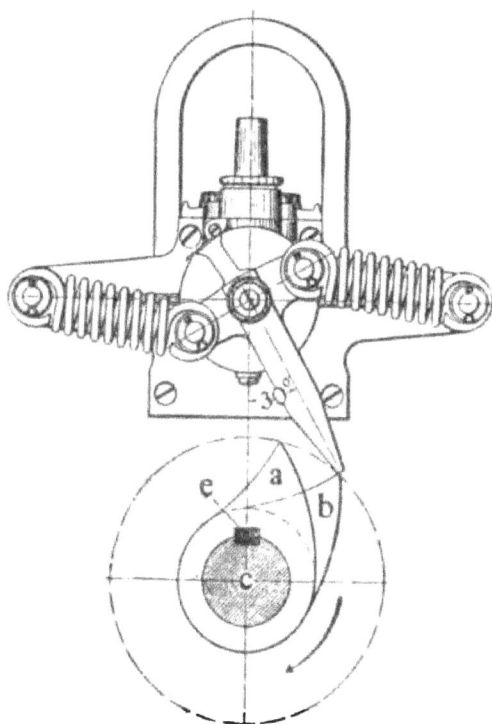

Fig. 5.—Bosch Oscillating High Tension Magneto.

Two cams, (a) and (b), are mounted side by side on the cam shaft,
one cam (a) giving a retarded spark for starting the engine, and the
other giving an advanced spark for normal running. The cams are
mounted on a sleeve which may be moved longitudinally on the shaft
(c), guided and being prevented from turning by the key-way or
feather (e). Thus by moving the sleeve on the shaft, the trip lever
may be either brought into contact with cams (a) or (b), A spiral
spring is placed between the cam (b) and a collar mounted on the

shaft which tends normally to keep the cam (b) in contact with the trip.

The circuit breaker is set so that the primary circuit is broken when the armature is moving at its greatest velocity so that the maximum spark may be obtained.

INDUCTOR MAGNETOS

An inductor type magneto has no revolving primary winding as in the shuttle type, the winding being mounted on a stationary core,

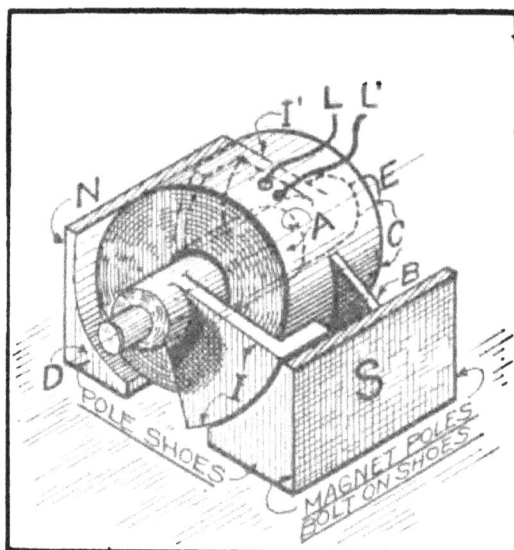

Fig. 6.—Diagrammatic View of Remy Inductor Type Magneto Generating System.

which is attached to the frame of the magneto. The only revolving parts in the inductor are the shaft and a sector shaped mass of iron or steel called the "inductor." With a stationary coil it is possible to dispense with brushes and collector rings and in most cases this construction permits of more room for the wire and insulation. At the present time, the Remy, Pittsfield K.W. and Dixie are the only inductor types on the market and the reader is requested to look for these types under their respective headings.

In Fig. 6 is shown a perspective of a Remy inductor magneto system in which N and S are the two magnet pole shoes and C is the primary winding. The iron inductors I and I^1 are mounted on the shaft D-E and revolve with it, alternately coming opposite the pole shoes N and S.

In the position shown, the magnetic flux from the pole N passes across and through the inductor I^1, through the iron shaft D-E and back to the pole shoe S through the inductor I. As the flux passes through the shaft it also passes through the coil of primary wire C.

As the inductors revolve they interrupt and reverse the flux passing through the coil, thus producing an alternating current in C. Consider the inductors turned through 180 degrees from the position shown. The inductor I will now be in contact with N, while inductor I^1 will be in magnetic contact with pole shoe S, thus reversing the direction of the flux through C. There are two reversals of flux per revolution and hence it is possible to have two sparks per revolution. The coil is supported from the pole shoes by the lugs B and F.

The ordinary type of horseshoe magnets are placed over and bolted to the pole shoes. The ends L-L^1 of the coil C lead directly to the transformer coil without the use of brushes or slip rings, making a very simple and reliable job.

True high-tension inductor magnetos are particularly desirable, since in the shuttle type it is always a matter of difficulty to properly place the high-tension brush and high-tension collector rings that are connected to the secondary winding. The K.W. and the "Dixie," described elsewhere in this book, are of the true high-tension type of inductor magnets, as the primary and secondary of the coil are mounted on the same core.

Inductor magnetos always deliver alternating current, as it would be almost impossible to properly mount a commutator.

REMY TRANSFORMER TYPE MAGNETO

In Fig. 7 the application of the Remy inductor principle is seen applied to the actual machine, the spark coil and the dash switch being omitted for the sake of simplicity. The front elevation is shown by the left view, while at the right is a longitudinal section taken along the center line of the machine. This is known to the trade as Model S.

Mounted on the driving shaft S108 are the two inductors S21 connected with the short sleeve S20. The primary coil is shown by S10 and is held in place by the clips S19. The breaker cam S157 is held on the shaft and works against the hardened steel plate S31 (see front view for circuit breaker). The hardened steel plate is fastened to the rocker arm S30, which also carries the contact spring S33 on which the platinum contact points S36 are mounted. A spiral spring S55 holds the rocker arm against the cam at all times. Screws or knurled nuts S54 and S48 are for the purpose of connecting the lead to the coil

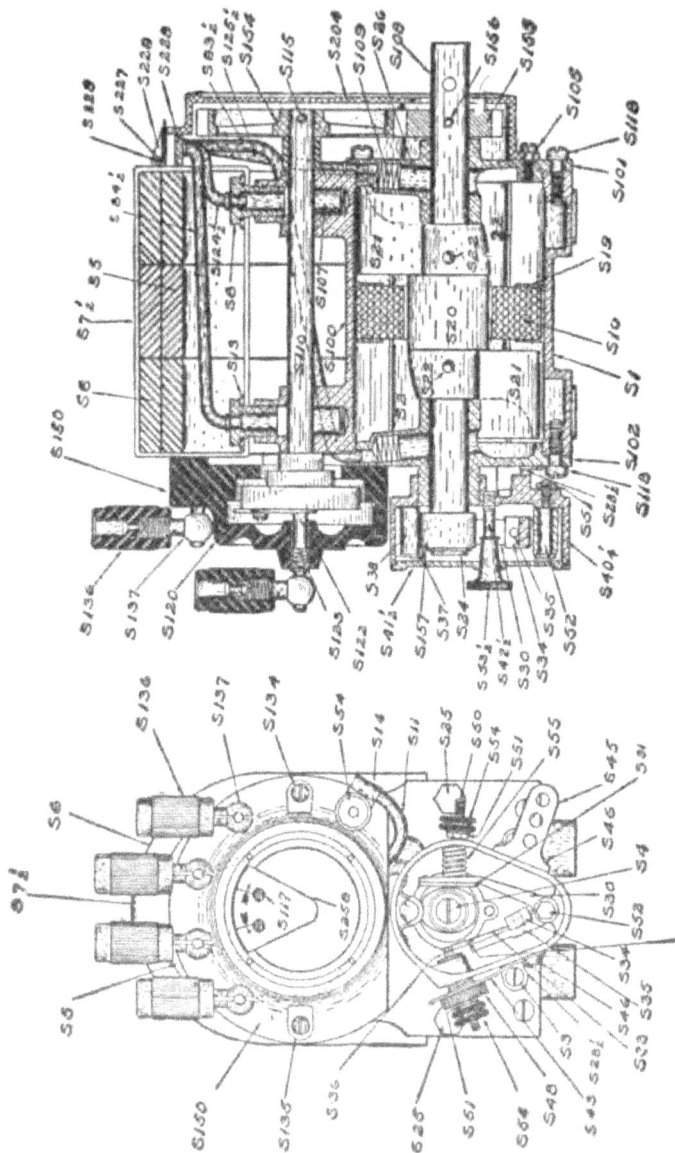

Front Elevation and Longitudinal Section Through
Remy Magneto.

and for adjusting the platinum points respectively. S45 is the timing lever, which moves the entire housing for the advance and retard.

Consulting the longitudinal section, the pinion S155 mounted on the inductor shaft meshes with the gear S154 which is mounted on the distributor shaft S110. This shaft is carried by two plain bearings at the front and rear which are fed with oil by the copper tubes shown entering the oil cup at the top and rear S228. The distributor disc is shown at the extreme left of the distributor shaft, which is made of some highly resisting material such as hard rubber. This is fed with high tension current from the brush S122 connected to the terminal plug, the brush being forced into position with the spring S123. With

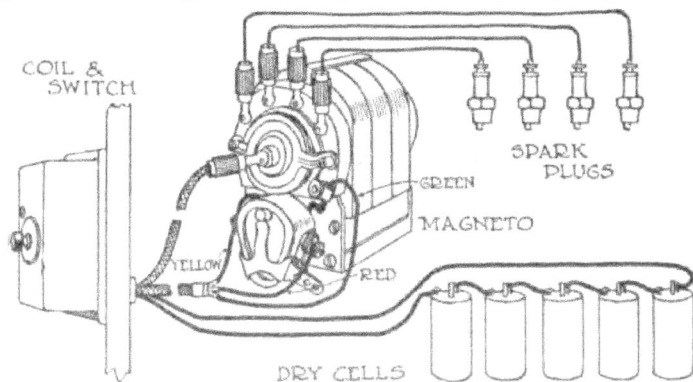

Fig. 8.—Circuit Diagram of Remy Transformer Magneto.
Courtesy "Motor Age."

a four cylinder motor the distributer shaft runs at half the speed of the inductor shaft.

Looking at the front elevation, the distributer disc is marked S258, and on the disc is shown the brass distributer segment S117 against which the brush from the high tension lead presses. As this sector revolves it comes in successive contact with the contacts of the four high tension plugs marked S136, a cable from each of these plugs running to the spark plugs in the cylinders. The contacts with which the sector meshes are in the form of four wires spaced equally around the circle and may be traced by following the dotted lines that run from the plugs to the distributer disc circle. The distributer parts are covered by a removable cap S120, which may be easily removed for inspection. This sector does not actually rub on the four contacts but comes within approximately 1/64 inch so that it is an easy matter for the spark to jump.

The proper points of connections for the coil leads are easily found from the accompanying circuit diagram which shows the colors of the leads and the points at which they connect.

"K. W." INDUCTOR TYPE MAGNETO

As with the Remy magneto, the primary winding of the K. W. inductor magneto occupies the space between the two revolving inductor masses, but unlike the former example this magneto gives four instead of two current impulses per revolution. The construction of the K. W. system is shown by Fig. 1 in which A and A^1 are the inductors and C is the primary winding. As the inductors are double ended and at right angles each inductor cuts the magnetic field four times per revolution, two times for each end.

This magneto may be used either as a low tension, low tension

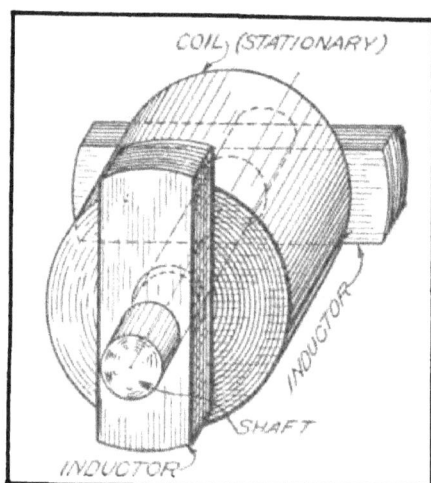

Fig. 9.—Inductors of K. W. Magneto.

transformer type, or as a true high tension magneto. When used as a true high tension type the usual circuit breaker and high tension distributer are mounted directly on the instrument. Usually the primary coil of the low tension K. W. magneto is made up of a strip of sheet copper rolled on an insulating spool, since this construction gives a lower resistance and hence a greater volume of current than with the round copper wire generally used, space for space.

Like all magnetos, the true high tension K. W. is positively driven from the engines through gears or chain, and as there are four impulses per revolution instead of two, the speed relative to the engine is half that given for the shuttle type armature. The greater number of impulses makes this magneto very desirable on eight and twelve cylinder motors as the rotative speeds are much reduced.

"K. W." HIGH TENSION MAGNETO

The "K. W." high tension magneto is of the inductor type and generates high tension current directly without the use of a spark coil. The arrangement of the coil and inductors is practically the same as in the case of the low tension K. W. magneto described under the head of "Inductor Magnetos," except for the fact that the generating coil carries both a primary and secondary winding. Using the same form of inductor with the two bars at right angles to one another, the high tension K. W. also gives four current impulses per revolution.

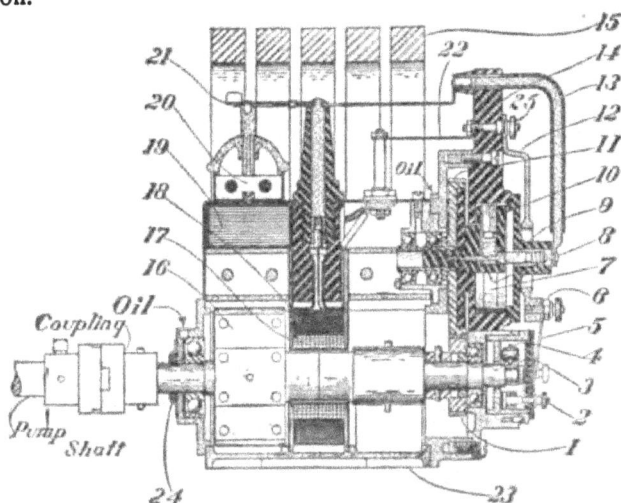

Fig. 10.—Longitudinal Section Through K. W. High Tension Magneto.

A longitudinal section is shown by Fig. 2 in which 16-16 are the inductors and 17-18 are the primary and secondary coils respectively. The two coils are wound on a single spool which occupies the space between the two inductors. A heavy hard rubber insulator is mounted on the spool, this carrying the high tension lead from the secondary coil to the point where it connects with the bridge 21. This is shown cross-hatched with heavy black lines. The current from the primary winding is led to the circuit breaker through the connectors 22, 25, and 12, the final connections coming from 12 to the terminal 6, and then through strip 5 to the breaker jaws.

High tension current from the bridge 21 splits two ways, one way being to the distributer through 13, and the other being to the safety sparks gap 20. Current enters the porcelain cap through a point, and if a sufficiently high voltage exists it jumps across the gap 20 to the point mounted on the condenser case 19, and thence to the frame and ground.

A condenser 19 is connected across the primary winding in the same way as in the case of the shuttle wound armature. As one end of the primary winding is grounded, one side of the condenser is also grounded, thus leaving only one wire to extend from the primary to the condenser. The free end of the primary winding is closed and broken by the interrupter contacts, a spark being produced at each interruption.

High tension current from the lead 13 enters the distributer by the way of the brush 9, and from there connects with the main distributer brush 10. As this distributer brush revolves it comes into successive

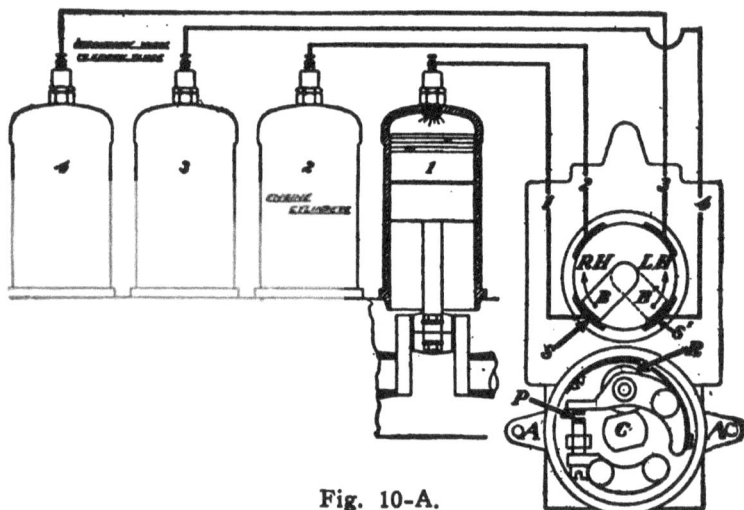

Fig. 10-A.

contact with the metal distributer segments 7 which are arranged around the insulating distributer block at evenly spaced intervals. The number of segments, of course, correspond with the number of spark plugs or cylinders.

The method of connecting the magneto distributer to the cylinders is shown by Fig. 10-A in which S and S¹ are two of the distributer segments, and B is the rotating distributer arm. High tension current is led to the cylinders 1-2-3-4 from the respective distributer terminals 1-2-3-4. The dotted distributer arm B¹ shows the position for left hand rotation while the full line arm is the position for right hand rotation..

A detail of the circuit breaker is also shown by the front elevation.

The inductor, or "Rotor" shaft as it is sometimes called, is mounted on ball bearings which require a minimum of attention and lubrication. The inductors are laminated, that is, are built up of sheets of soft iron or steel to increase the magnetic effect in the coil. It will also be noted that a sleeve is placed over the shaft for the purpose of increasing the area of the core that passes through the center of the coil and between the two inductors.

Unlike the case with the majority of magnetos, there are five magnets used, each magnet being of a square cross-section. This gives a very powerful field and an intense spark at extremely low speeds.

A flexible coupling is shown at the left end of the shaft which prevents strains from being thrown into the magneto bearings should the magneto and driving shafts be thrown out of alignment. As shown, the magneto is driven from the extended end of the pump shaft.

HERZ (RUTHARDT) MAGNETO

Probably one of the unusual features of this interesting magneto is the construction of the magnets. There are thin circular steel discs having the opening for the armature bored at the lower edge to the

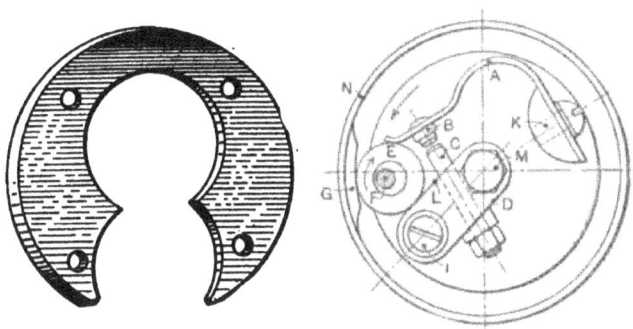

Fig. 11.—Magnet and Circuit Breaker of Herz (Ruthardt) Magneto, Showing One-Piece Field and Pole Shoes.

exact size of the tunnel, no cast pole pieces being used. The complete magnet is built up by stacking ten or twelve of these discs in a row and then riveting them together with four longitudinal rods. One of these discs is shown by Fig. 11.

After machining the bore for the armature, the interior of the bore is polished so as to obtain an exceedingly small air gap between the armature and the pole. This gap in the Herz Magneto is only 0.0015 inch which allows of a maximum magnetic flux through the armature. By avoiding the use of pole pieces the magnetism is again conserved since considerable leakage takes place at this point in the ordinary type.

The armature is of the usual shuttle type with a primary and secondary winding, the primary circuit being interrupted in the usual way by a circuit breaker. In this instrument the secondary is not a continuation of the primary as in other high tension types, but is thoroughly insulated from the primary as well as from the metal of the frame.

A feature of the circuit breaker is that it has no bearings and therefore requires no oil and wearing parts are avoided in it. An insulated block D is secured to the rotating disc of the breaker box by the screw I. A spring A with a platinum contact B is fastened to K. Screw C holding a platinum point is the adjustment for the breaker and is threaded into the block D.

The edge L of the block D forms a guide for the fiber roller E which has considerable play around the small spindle F while retained by a flat spring not shown. The end of the spring A rests against the roller pressing it forward, this action being aided by the centrifugal force. The roller which runs on the interior of the rim N will allow contact between points B and C except when it passes over the cam-face G. This pushes the roller inwards, forces the spring A outwards and separates the points. The number of cam-faces G used, depend on the number of cylinders in the motor.

The rim N which is the outer shell of the breaker box may be rocked to and fro to obtain the retard and advance. The bolt M conducts the current from the armature to the contact block D. A spring in the cover of the breaker box rests on M, terminating in a binding post which carries the wire from the cut-out switch. This spring and binding post are, of course, insulated from the frame of the magneto.

PART VII

MAGNETOS WITH SIMULTANEOUS ADVANCE OF BREAKER AND MAGNETIC FIELD

MEA MAGNETO

In the usual type of magneto the magnets and pole shoes are stationary while the circuit breaker housing is capable of being rocked back and forth to obtain the desired advance and retard of the spark. Since a maximum voltage is obtained only at one position of the armature in the magnetic field it is evident that the circuit breaker opens the primary circuit at many points where the voltage wave is comparatively low, while the housing is being moved from advance to retard. In other words, it is possible to obtain the maximum spark at only one position of the breaker housing.

In the majority of cases the breaker is set for the maximum spark with the housing in the fully advanced position. This, of course, causes the weakest spark to occur at full retard, the very point at which the strongest spark is required, as the retarded spark is always used in starting. With a low cranking speed, a poor mixture and a cold motor the already weak retarded spark is still further reduced in igniting value, often making the starting operation very difficult.

When running the engine at low speeds or in cases where the motor is overloaded, as in climbing hills, the effects of the retard are again a cause of annoyance, the failure of the spark to ignite the charge often causing the engine to stall at a critical time. Attempts, therefore, have been made to obtain a uniform spark independent of the timing.

With the Mea magneto the magnets and breaker housing practically are in one piece, the breaker therefore always opening the primary circuit at a constant armature position. The advance and retard are obtained by rocking both the magnets and the circuit breaker as one unit, the relative positions of the magnets and breaker remaining constant. This result is obtained by mounting the entire magneto on a rocking cradle in such a way that the relation between the armature and pole pieces is changed when advance or retard of the spark is desired. With this method of mounting there is no limit to the angle of timing, since with a suitable supporting cradle it would be possible to rotate the entire frame through an angle of 360 degrees without changing the quality of the spark. Practically the Mea has a timing range of from 45 to 70 degrees, since this is more than ample for any ordinary condition met with in automobile work.

To obtain the greatest benefit from the rocking cradle and to have a low compact instrument, the Mea magnets are bell shaped, with the magnet legs lying parallel with the armature shaft. This can be seen from Fig. 6 and Fig. 7, the elevation showing the magnets as a horizontal cylindrical frame extending from left to right. In the section the magnets are shown by the figure 100.

Fig. 1.—Elevation of Mea Magneto.

Consulting the cross-section, Fig. 7, it will be seen that the armature 1 is of the conventional shuttle wound type, the condenser 12 being contained within the armature. The high-tension lead from the secondary winding connects with the high-tension collector ring 4, while the primary current is led to the circuit breaker through the hollow right-hand shaft by the connector bar 24. The armature is carried by the ball bearings 17 and 18.

The magnets 100, indicated throughout by the double cross-hatch lines, are fastened to the frame 61 at the left, the latter carrying the ball bearings. The frame 61 is mounted in the cradle 53, in which it is free to rock to and fro for the advance and retard. At the right a similar bearing is formed in the cradle, at 60b, the parts 60b at the right and 60a at the left being the upper halves of the cradle bearings. Two spring controlled plungers 111 at right and left press against the rocking frame to prevent it from slipping loosely in the bearings. The magnetic pole shoes are the thin horizontal strips shown directly opposite the armature 1.

Primary current from the armature led through the conductor 24 to the circuit breaker plate 28, which is insulated from the frame. On this plate is mounted a spring 30, which carries the platinum contact point 34. This is the movable contact point. A fiber cam roller 31, which revolves with the plate 28, rests against the cam plate 40 at the back. The cam plate being provided with two cam projections causes the cam roller 31 to strike the spring 30 twice per revolution,

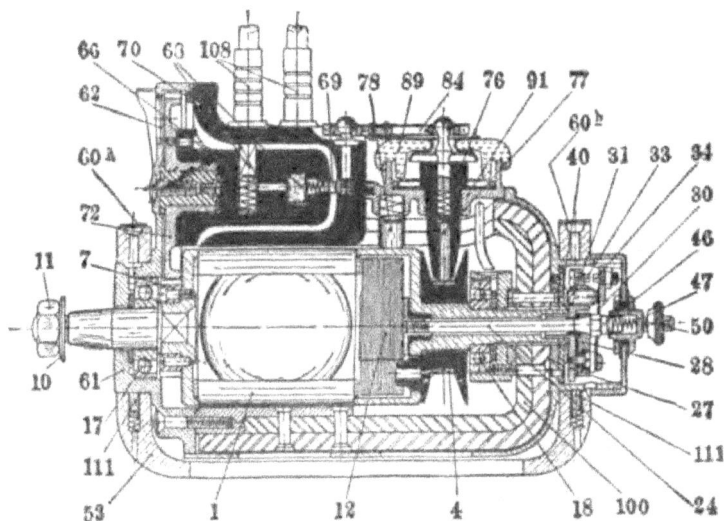

Fig. 2.—Section Through Mea Magneto.

breaking the contact between platinum points 33 and 34 and opening the primary circuit. The point 33 is grounded to the frame through the plate 27.

As the inner end of the primary winding is grounded to the armature core, the return current from the plate 27 flowing through the frame passes through the grounding brush 78 on its way to the armature. This completes the primary circuit.

The breaker housing is closed by a cover 74, which supports a carbon brush 46 through the insulating brush holder 47, this brush coming in contact with the primary current connector 24. The terminal screw 50 connects with the brush, and a lead from this terminal runs to a short circuiting switch on the dashboard. When the switch is closed, the primary current from the armature passing through terminal 50 is grounded, which of course will stop the generation of current and will therefore stop the engine.

High tension current from the secondary winding is taken from the high-tension collector ring 4 by the brush 77. This brush is supported by the insulating holder 76, a spring in the holder being used to force the brush against the collector ring. The holder 76 and the low-tension brush 78 are fastened to the plate 91, which in turn is supported by the magnet frame. Current from the high-tension brush is led through the bridge 84 to the distributer brush 69. The connection from the brush holder 76 passes through the insulating cover 89, which also acts as a safety spark gap.

The distributer consists of an insulating cover (stationary) 70 and a rotating part 66, the latter being driven from the armature shaft by the steel and bronze gears 7 and 72. Current reaches the rotating part of the distributer through the brush 69. From here the current reaches the two distributing brushes 68, which are placed in slots cut in the insulating material, the brushes being placed at right angles to one another. These brushes make alternate contact with four contact plates which are imbedded in the insulation of the stationary part of the distributer. These contact plates are connected to the four high-tension spark-plug terminals 108, which are located on top of the distributer.

The relative location of these plugs can be seen from the side elevation of the instrument, where it will be noted that the plug connections are arranged in two rows of two plugs per row. In the front of the distributer are small windows behind which appear numbers that are engraved on the distributer gear. These numbers correspond to the number of the cylinder that the magneto is firing at that time. This makes the reconnection of the magneto a simple matter.

A frame or cradle 53 carries the magnets, distributer, armature, etc., in such a way that the magnets, and with them the timer, can be rocked back and forth by a timing lever that is mounted on one side of the magnets. The spark is advanced by turning the magnets in a direction opposite to the direction of armature rotation.

The one-cylinder magnetos are similar to the four-cylinder except that the distributer is omitted.

DIXIE HIGH TENSION MAGNETO

The "Dixie" magneto is of the high tension type, but is entirely different in construction from any magneto on the market. It is of the "inductor type," that is, there are no revolving windings or wire, the only revolving part being an iron mass that directs and breaks the magnetic field in a stationary coil. Figs. 4-5-6-7 show the cycle of operations of the revolving inductor marked N-S, while Fig. 8 shows the arrangement of the circuit.

Unlike the usual magneto, the shaft is at right angles to the plane of the "U" form magnets, as shown by Fig. 12, in which M is the magnet, C is the core of winding, and W is the winding which consists of a primary and second coil. The two shoes of the inductor, N and S, are practically at all times in contact with the ends of the horseshoe

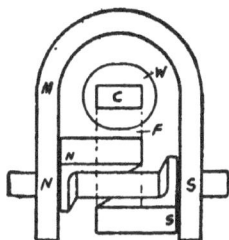

Fig. 4

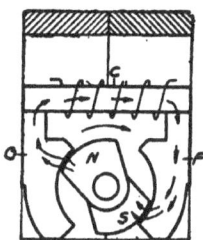

Fig. 5

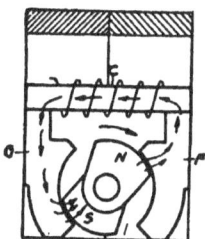

Fig. 6

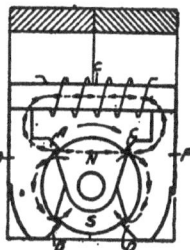

Fig. 7

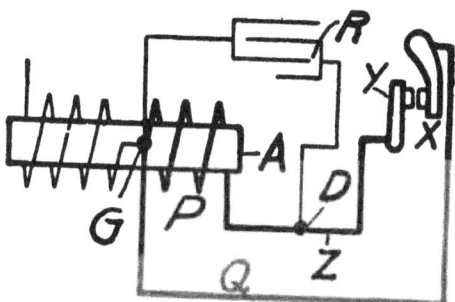

Figs. 4-5-6-7-8.—Construction and Circuit of "Dixie" High Tension Magneto. Inductor Type With Advance and Retard of Field.

magnet, and consequently at all times have the same polarity. Looking at the view shown by Fig. 5, which is taken at right angles to the first figure, we see the sector-shaped shoes of the inductor N-S.

Since the two revolving shoes are always of the same polarity, they alternately produce opposite polarities in the core poles G and F, causing a reversal of magnetic flow through the iron core C of the winding. At each reversal a current impulse is induced in the stationary winding. In the position shown the current flows to the left from the upper shoe N.

A turn of a little less than 90 degrees brings the condition shown in Fig. 6, where the shoe end N comes opposite the core pole F, causing a flow of flux to the right. In Fig. 7 the shoes are shown in the mid-position, short-circuiting the core C and thus cutting the magnetic flux flow to zero.

This advance from zero magnetic flux, and back to zero, causes the maximum current in the winding, since the magnetic variations are sharp and sudden. It should be understood that the current depends on the **rate** or **speed** with which the magnetic variations take place, and not altogether upon the intensity of the magnetic field.

Since no moving wire is used, the difficulties ordinarily experienced with the shuttle type armature in collecting the current are removed, having less parts to wear and avoiding unnecessary friction and brush trouble. There is more space for the windings and their insulation, and the wire is at no time under the strain caused by centrifugal force.

Fig. 8 shows the circuit diagram of the "Dixie," in which the heavy winding P is the primary, A is the iron core and G is the fine wire secondary. One end of the condenser R is connected at the point G, where the secondary winding connects to the end of the primary winding, the other end of the condenser being connected to the outer end of the primary, thus placing the condenser in "shunt" or "parallel" with the primary winding. The interruption of the primary circuit is performed by the circuit breaker contacts X and Y, which are in series with the primary coil P. The end of the secondary wire at the left connects with the revolving arm of the high tension distributer.

The coil is mounted in connection with the circuit breaker in such a way that they move back and forth together for advance and retard, thus causing the same intensity of spark for any position of the spark lever. In this magneto the circuit breaker points and the magnetic field always retain the same relative position.

INSTALLING HIGH TENSION MAGNETOS

The installation of a high tension magneto requires a considerable knowledge of machine work and of the functioning of the gasoline engine, especially when installing the instrument on an old car or on an engine not built especially for magneto ignition. The following instructions are therefore not intended for the car owner or chauffeur but for the automobile mechanic or garage manager who are provided with the necessary experience and shop facilities for making the extensive alterations.

When the matter of installing is simply one of replacing an old magneto with new, on an engine already equipped for magneto ignition, the owner or chauffeur can often make the installation by consulting the chapter on magneto timing, starting with Page 213.

Probably the first point to consider is the gear ratio existing between the armature shaft of the magneto and the crank-shaft of the motor. This depends entirely upon the number of cylinders as before explained, the magneto of a four-cylinder engine running exactly at crank shaft speed while a magneto for a six-cylinder motor runs at exactly 1½ times crank shaft speed. For any other number of cylinders consult the table on Page 73. It should be remembered in this connection that the high tension magneto must be **positively** driven from the engine by either a chain or gears so that there is no slip between the two, and also, that the gear ratios must be **exactly** as given for the different number of cylinders. A belt or friction drive cannot be used.

In ordering the magneto give the number of cylinders so that the distributer will be correctly arranged, and also give the direction of rotation of the armature shaft. If the engine is for other than an automobile give the bore and stroke of the cylinders and the speed of rotation. If for a car equipped for magneto ignition give the name and model or year. Some of the more modern magnetos are equipped so that they can be used with either right hand or left hand rotation by a simple adjustment of the distributer disc, while others must be adjusted for rotation direction at the factory. Give location of timing arm at right or left.

In regard to direction of rotation first determine the method by which you intend to drive the magneto. Whether from the crank-shaft or cam-shaft and also note whether it will be necessary to introduce intermediate gears between these shafts and the magneto. It should be remembered that an intermediate gear reverses the direction of rotation. After this matter is settled trace out from the crank shaft rotation, the direction of rotation of the magneto following through all the proposed intermediate gears. If the magneto is driven from the cam-shaft through a single pair of gears, the magneto rotation will be opposite to that of the cam shaft. Direction is taken when facing driving shaft end of magneto.

The magneto should be mounted directly on the engine frame by means of a metal plate or bracket and rigidly bolted in place. The magneto base must be in absolute metallic connection with the engine frame without an intervening coat of paint or layer of dirt since it is through this point that the grounded current returns from the spark plugs to the magneto. Never use paper or wooden shims under the magneto. Do not place the magneto against a hot exhaust pipe or in any place where the temperature is likely to rise above 150 degrees. Take care, also, not to have the magneto breaker box located near the carburetor or gas pipes since the spark at the breaker is likely to cause ignition at points of leaks or overflow.

A flexible coupling must be placed between the driving shaft of the magneto and the driving shaft of the motor, preferably of the type known as the "Oldham." This prevents strains from being thrown into the delicate magneto armature through a lack of perfect alignment of the magneto and engine driving shafts. These couplings can be obtained, usually, from the maker of the magneto and permit of a considerable degree of error in the lining of the two shafts.

Gearing should be made as accurately as possible with little ba^klash between the teeth or play in the bearings. Any great amount of play or lost motion between the crank and magneto shafts is likely to cause annoying intermittent changes in the timing of the ignition. If chain is to be used the flexible coupling can be omitted, but care must be taken not to have the chain too slack or too tight. The form will have the same effect as lost motion in the gears, while the latter will of course cause unnecessary wear in the magneto bearings.

If possible, place the magneto in such a position that the primary circuit breaker and distributer can be easily inspected for cleaning and adjustment. Be sure that the oil holes are accessible, and that the distributer is in such a position that cables leading to the spark plug are not unnecessarily long or do not trail over the hot exhaust pipes. Every additional foot of high tension cable gives an additional chance of leakage and a weak spark at the plugs. Note the position of your spark control lever and order the magneto so that the timing lever on the breaker box of the magneto is on the required side of the box.

On all engines exposed to the weather, especially with tractors and farm engines, provide a watertight cover for the magneto. While all magnetos are specified as "Waterproof," it certainly does no harm to provide additional protection. Oil and grease are especially to be provided against for oil is most detrimental to the insulation and high tension cables. Seek to avoid placing magneto near the rubber connections used on the cooling system for a leak may cause serious trouble.

Having disposed of the subject of location and having the gears and magneto attachments at hand we will now consider the mounting and wiring.

Both the gears and couplings to the magneto should be keyed firmly to their respective shafts, and no dependence should be placed on set screws or wedges since both of these makeshifts are bound to slip in time and destroy the accuracy of the timing. Temporarily, or at least during the period when the magneto is being timed, it is often convenient to fasten the couplings with a set screw, but the key ways should be marked and cut at the earliest possible moment thereafter.

Bolt the magneto firmly on the bracket, and place the new gears and couplings in place, and key the gears that are to drive the shafts. Idle gears that revolve on the shafts should be carefully bushed with either bronze or cast iron and provided with means of lubrication. Chains and gears should be thoroughly protected against dirt or grit by suitable metal housings. Only hardened steel gears, accurately cut, are suitable for the purpose, and should have a sufficiently wide face to insure long service and a minimum amount of noise. A face ½ inch wide should be the least width of face and should preferably be ¾ inch. See that the magneto and engine driving shafts line up and then turn engine over slowly by hand to see if there is any tendency to crank or bind. There should be a slight clearance between the flanges of the coupling to avoid any possibility of binding due to lack of truth in the mounting of the couplings.

Now loosen the temporary set screw in the coupling on the magneto side so that the armature shaft can be turned back and forth in relation to the driving shaft for the timing operation.

The wiring depends entirely upon the type of magneto used, or whether it is of the single, dual, duplex or two point type. Diagrams showing the principles involved in these different types are shown on Page 92, and are described in detail in Part V starting on Page 91. The wiring of the true high tension type is also different from the transformer type of magneto as previously described. Even among magnetos of the same type there are differences in the actual connections made between the coil, magneto and switch. Specific wiring diagrams for the Bosch single, Remy transformer, and K. W. magnetos are given in this chapter, with diagrams on pages 94, 99, 107, 110 and 117.

Wiring diagams of all true high tension type magnetos without battery auxiliary are wired in practically the same way as shown by Fig. 1 on Page 92 or by the Bosch diagram on Page 99; that is, a separate lead from the distributer to each spark plug, and a low tension lead from the breaker box to the switch on the dash-board. From the remaining connection post of the switch a ground wire runs to "ground" or rather to the frame of the car or engine. When Battery auxiliaries are used or with transformer type magnetos, there are additional wires between the magneto and spark coil and between the coil and battery depending on the magneto type. The number of these wires in any system can be taken from the diagrams on Page 92 corresponding to the type used.

For the present, however, we are concerned with the number of wires, their direction of running, and the method of their support rather than the actual points of connection on the magneto and coil. When running the wires leave the ends long enough so that they will reach to any binding post on the instrument. The makers either have colored wires on the cables that connect with terminals of corresponding color on the instruments or number the connection points on the spark coil that are to be connected with corresponding numbers on the magneto. The ground wire to frame is marked "G" and the Battery connection is marked "B" when not fully written out.

All wiring, whether high or low tension, should be firmly supported by insulating supports and should be of the shortest possible length. They should not come into contact with the frame or metal parts and most emphatically should not come into contact with sharp edges or corners that will be likely to chafe through the insulation through the vibration of the engine. They should be readily accessible for repairs or renewal and protected from an excess of grease or oil. This applies particularly to the high tension cables, for oil is most effective in the destruction of the rubber insulation.

All high tension wires should be of the very best grade of specially made high tension cable. The best is none too good for withstanding a voltage of from 10,000 to 25,000 volts. All wiring, whether low or high tension should be of the cable or stranded type, built of a cable of fine wires. A cable is easier to handle than a solid wire and is much less likely to break or strip. No less than a No. 14 B and S gauge wire should be used, for smaller wires are mechanically weak and most likely will be a source of trouble. The best possible method of supporting wires is by means of fiber tubes, through which they are run to the point of connection.

Different methods of supporting the high tension wires running from the distributer to the spark plugs is shown by Fig. 12 on Page 35. The neatest and most desirable methods are those shown by the two lower cuts where the wires are run through tubes, the leads to the plugs leaving at the side of the tube. The magneto distributed is shown at the left of the diagrams.

In regard to the gear ratios given in the table on Page 73 it should be mentioned that the relative speeds are only true for four cycle motors. For two cycle motors the relative speeds for **more than two cylinders** will be double those given in the table. This is due to the fact that a two cycle motor cylinder gives twice as many impulses in a given time as a four cycle.

With the magneto, coil and wiring installed we are now ready to time the magneto in regard to the piston position in the cylinders. This most important part of the installation will be taken up in detail in the latter part of Part XI under the sub-head "Timing Magnetos."

PART VIII

IGNITION TROUBLES—MAGNETO AND BATTERY SYSTEMS

REPAIR—ADJUSTMENT—CARE
OF
HIGH AND LOW TENSION APPARATUS

This chapter is devoted to the repair and adjustment of the ignition system by the simple means open to the operator of automobiles and gas engines, and covers the routine of attention necessary for the adjustment and care of high tension magnetos.

If electrical matters are not your forte, the ignition system is a most likely point of failure; hence the first thing to do when the engine unexpectedly refuses to give the regular beat of its explosion, (the skip is quickly noticed) is to see whether you obtain a spark between the frame of the machine and the insulating cap of each sparking plug when the engine is turned by means of the starting handle through one or two complete revolutions (of course with mixture cut off and compression released where provision for this is made in the machine).

When doing this connect one end of the copper of a thickly insulated wire to the frame of the machine and place the other end very close to the exposed binding screw or brass cap of the spark plug; if you get a good spark you may be almost sure that the ignition is not the cause of the stoppage. If you cannot get a piece of insulated wire, a key or a spanner will do as well, provided you are careful to avoid getting a shock by keeping the metal of the spanner well grounded— that is, in contact with the frame. If you get either a fat spark or a strong shock, the ignition is probably all right. If you greatly dread a shock, undo the high-tension wire, and, holding it by the rubber insulation, try the high-tension spark, which should be about ½ inch long in air and very bright. Special terminals are made for the rapid disconnection of the high-tension wire for this purpose. There are, however, five cases in which you will obtain a spark or a shock, although the ignition is the cause of failure.

The first case is when the battery is practically empty, but has had time to recover in a temporary manner owing to the stoppage of the car. You will probably find that the engine will re-start and only run a little way. The only cure for this is to have a second battery, which should be switched on with a two-way switch.

The second case is when some one of the wires of the system is making a bad connection or is partially broken. In this case the running of the car will shake the broken or loose parts asunder, whereas the stationary car may give you a satisfactory spark. The quickest cure for this in the long run is not to hunt for faults, but to tighten one by one every binding screw and turn the handle again. If this does not cure it, pull out one by one every wire in the car and replace them with your spare wires. The reason for not hunting out the fault is that the break has probably occurred within the thickness of the insulation itself, and is not discoverable without instruments or, at any rate, a tedious search.

The third case is when the two points of the spark gap within the cylinder are too far apart, or have too much oil, soot or moisture on them to allow a good fat spark to occur within the cylinder. Remember that a spark does not occur with the same ease in the compressed gas of the cylinder as it does in the open air.

Fourth.—It is possible that the porcelain or mica insulator of the sparking plug is cracked or allows the current to flow through it. This is cured at once by inserting a spare plug.

Fifth.—If the high-pressure spark appears to be thin and weedy, so that you suspect it of not being hot enough to ignite the gas within the cylinder when under compression, it is probable that one of five things is the cause—

1. The battery has run down.

2. The trembler blade and its platinum contacts are out of adjustment.

3. There is grease or oil upon the make-and-break spring, if a non-trembler coil is in use.

4. The condenser has somehow become disconnected or punctured.

5. The high-tension windings on the coil or the high-tension wire have partly broken down their insulation.

Therefore, switch over to your spare battery; clean the platinum points of your spring contact by rubbing a visiting card between them or even by smoothing them out with a very fine file and removing all filings very carefully. The condenser trouble is not curable on the road, but you may be able to run your engine slowly home if you allow the compression cock to leak a little. If touring in remote parts it is worth while to carry a spare coil or even to have a complete stand-by ignition.

Damp.—Water is a conductor of electricity; therefore the

porcelain plug should be wiped after a damp or foggy run. The rubber-covered wires where they approach the terminals should also be kept dry. This will be found impossible if there is any exposed braid or tape to get damp and to collect and retain it. Therefore, remove the braid and tape from the end of all wires for a length of about an inch close to the terminals. Do not remove the braid from the entire length of the wire, because it is a useful protection against breaking and fretting, but dip the whole of the braided wire before using it into a bath of melted paraffin wax, wiping off the surplus.

Supposing that neither a spark nor a shock is obtained at the spark plug of the cylinder which is misfiring, it is generally safe to look in turn for one of the following defects:

(i.) The battery completely exhausted (try it with a voltmeter which takes the full normal current, say 1 ampere).

(ii.) There is a disconnection or a break in a wire—examine first the high-tension wire as being the simpler. See also that the switch has not accidentally been moved to the "off" position. See that the lead lug in the battery is not broken.

(iii.) The platinum tip has fallen off from the trembler blade or from the spring contact, or from the platinum-tipped screw; or the entire trembler blade has come loose from its clamp.

(iv.) One of the coils has completely broken down. If so, the click of a spark is generally audible by putting one's ear close to the coil-box.

In applying these various suggestions the reader is credited with a certain amount of acumen. Thus he will at once surmise that if only one of four coils fails to give a good spark it clearly cannot be the battery that has failed; indeed, trouble in this case is not to be expected in any part which is common to the whole four coils, as, for example, the ground wire, or the switch, or the commutator, save at the one contact corresponding to the one cylinder showing the faulty ignition.

Cleaning Sooty Plugs.

With a two, three or four-cylinder motor, it is quite possible to clean a sooty ignition plug without removing it from the cylinder. The "modus operandi" is as follows: Detach the high-tension wire from the misbehaving plug, open the compression cock of its cylinder, and run the engine on the other cylinders. Then hold the terminal of the detached wire, being very careful not to touch the metal part with the fingers, a very short distance off the end of the plug, so that the spark jumps to the latter. The wire should be held by the insulated part at least two inches from the bared terminal. At first the cylinder will be heard to be missing, but very quickly the reverse will be the case, and the dirty plug will be found to have cleaned itself, as to all intents and purposes a spark gap is established. Then switch off, shut your compression cock, attach your wire again, and start up.

How to True Up the Contact Screw.

In devising a jig for trembler screw points, the simplest possible form, if there is no difficulty in getting either a "tap" which will fit the trembler screw, or in cutting a tap to do so, all that is necessary for the trembler screw device is to drill and tap a hole through a piece of ⅜-inch steel plate, ground perfectly true on one face, and then harden the plate. The hard, flat face acts as a guide for the file, insuring that it travels truly in the same plane, and the fact that the trembler screw is held by its own thread is a guarantee that the face of the point is at right angles to the line of the screw. This way of doing the point is the simplest, but supposing the trembler screw cannot have its thread matched without trouble and expense, the jig illustrated in Fig. 1 obviates any difficulty in this direction. It consists simply of a piece of cast steel bar, bent round as shown, and having the face marked D ground quite flat. Through the center of this portion of the bent steel a hole is drilled, which is exactly the size of the outside of the trembler screw, so that it will

just push in from the under side easily. In the illustration
a slice has been cut out in front of the trembler screw A so
that it can be seen. Exactly opposite this hole in the other
arm of the bend, a second hole is drilled and tapped with any
convenient thread. Through this hole the round-ended set
screw B is inserted, its rounded end bearing beneath the
milled end of the trembler screw. The steel bend and the
screw B should both be hardened. To use this jig the

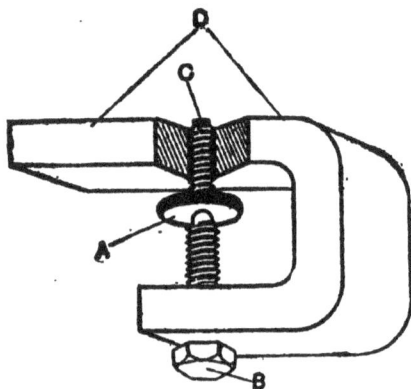

Fig. 1.
Jig for filing trembler screw points correctly.
A, trembler screw.   C, platinum point.
B, adjusting screw acting as a stop.   D, hardened flat face.

trembler screw is inserted, and the lower set screw B run up
until the contact point C can just be seen, on glancing along
the face D, to be sufficiently above that face to clean up quite
flat. A fine flat file then steadied along the true face D will
complete the operation, and a perfectly level surface will be
obtained for the point C.

Truing Up the Blade Contact.

The jig for the trembler blades illustrated in Fig. 2 consists
of a cruciform base, the central portion of which is bored and
screwed to receive the circular table A. This table, which is
shown separately in the left-hand corner with its shank in
section, is made with a buttress thread as shown, and preferably its edges should be knurled or milled. Along one diameter of the base, and equidistant from the center of the table

A, are the screwed studs D D, which pass through holes in the two clips C C. These clips are as shown, and consist simply of two metal strips about ¾ inch wide and $\frac{3}{16}$ inch thick, bent over at the one end, and ground flat on the down-turned face opposed to the top of the table A at the other. Along the other diameter the studs or stops B B are fixed, these being screwed into the base and made from hard steel. The upper faces of these studs B B must be perfectly level, and exactly the same height from the base plate.

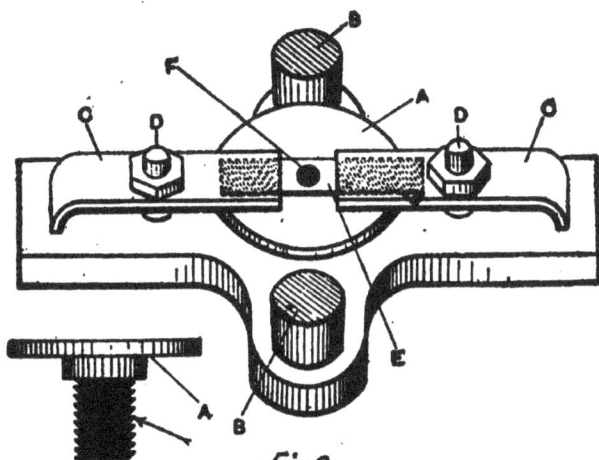

Fig. 2.

Perspective plan view of jig for filing the points of trembler blades flat.

A, adjustable table with buttress threaded shank.
B B, hard steel studs for grinding the file.
C C, clips holding blade to table.

D D, studs and nuts for tightening clips.
E, trembler blade.
F, platinum point.

The method of procedure is as follows: The trembler blade E is laid down on the table A with the platinum point F upwards. The clips C C are slipped over, so that the trembler blade is between the lower faces of the clips and the table top, but the nuts on the studs D D are left quite slack. A straight-edge is then held across the faces B B, and the table screwed up until the point F comes into contact with it. The straight-edge is removed and the table taken up just a shade more, according to how much has to be removed from the platinum point F. The nuts on the studs D D are then tightened down gently, when the blade is gripped firmly against

9

the surface of the table; and, moreover, since the table has a buttress-threaded shank, the pressure locks the table very securely in position. A fine file is then run across the studs B B, with the result that the point F is made quite flat and true with the blade in both directions.

Two steel plates, identical in shape, which can be clamped together by a pair of thumbscrews, and having a ⅛-inch hole drilled completely through the two, make an excellent jig for holding the end of the trembler blade when filing the hole to secure coincident setting of the points, this obviating straining or bending.

Adjusting Contact Breaker Screws.

In adjusting the contact breaker screws when the contact breaker is of the positive make-and-break type, care should always be taken to see that the small locking screw, which is provided in the split end of the screw-supporting pillar, is properly locked after the adjustment is complete. Also, it should be noticed whether locking up this screw affects the adjustment of the contact screw, as this sometimes happens. When the contact-making screw is not properly locked up, the constant tapping on it of the trembler blade invariably works it farther back, so that the adjustment does not keep correct for any length of time, and, consequently, the annoyance of misfiring is experienced very frequently.

A Cause of Irregular Firing.

Automobilists who have had experience with the old De Dion type of contact maker have at some time or other been troubled with irregular firing of the cylinder charge. This has been put down to various causes, chief of which, no doubt, has been a dirty contact between the platinum-pointed adjusting screw and the trembler blade. Other causes are loose contacts, either at the plug wires or in the primary circuit, dirty sparking plugs, nearly run-down battery, or a bad mixture.

A not infrequent cause of trouble, however, is due to the fact of the hole in the insulating quadrant of the contact maker itself wearing slack or oval. The constant knocking action of the notched cam on the trembler V-piece in time causes an oval hole in the quadrant, and consequently the correct action of the trembler is interfered with, and frequent adjustments of the screw become absolutely necessary. When the quadrant wears oval, it can be repaired by bushing it with gunmetal, but probably a more satisfactory method would be to fit a new contact maker.

A more lasting quadrant would be one made entirely of metal, a good hard gunmetal for preference, in which case the adjusting screw terminal and connections would have to be insulated by means of wood fiber or ebonite washers, and there would be no necessity for a ground wire to be the trembler terminal, as it would be constantly grounded through the metal quadrant. This has been employed in certain cases, and has given every satisfaction.

Trembler Contact Treatment.

When coil tremblers are behaving badly and require filing, it may happen that a suitable file is not at hand. Then a fairly good new surface can be obtained on the platinum end of the screw by tapping it lightly with a small smooth-faced hammer.

Loose Contacts and Faulty Firing.

A case is reported of a somewhat curious source of trouble in connection with a two-cylinder engine. The engine suddenly refused to fire, and after much trouble and delay in testing wires, etc., the fault was found to be ascribable to a loose platinum tip in the trembler adjusting screw. At first sight the platinum showed about one-sixteenth of an inch clear of the end of the screw, but when tapped with a hammer it disappeared entirely within the screw. The trouble disappeared altogether when a new screw which had its platinum point quite firm was fitted.

Trembler Fatigue.

It is pretty certain that ignition troubles are occasionally caused by trembler fatigue. A case in point which once occurred to a motoring authority would go to prove this. His engine would drive for from forty to fifty miles perfectly, running the car up all ordinary grades on its fourth speed; but shortly after that distance had been covered, one of the cylinders would begin to fire irregularly, and nothing would improve it. Coil trouble was suggested, but he was loth to believe this was at the root of the evil. The erring cylinder would not drive the crankshaft alone, when the tremblers of the other three cylinders were prevented from vibrating, although either of these three would perform fairly well by itself. No sooner was a new trembler fitted than the aforesaid weak cylinder ran merrily by itself, and the owner was moved to replace the remaining three old tremblers with new ones. These replacements greatly improved the firing and pull of all the cylinders, so that he was forced to believe that the old tremblers, which had been in constant use for eight months, wanted a rest.

Pitting of Trembler Contacts.

An annoying trouble, which in many cases is only a too frequent occurrence, is the pitting of the platinum contact on the trembler blade on high-speed trembler coils. The coil will sometimes work fairly well at slow speeds, but gives frequent missing when the engine is run at its normal rate, or accelerated. On examination of the platinum points mentioned, it is found that the screw appears to have its end melted in the form of a rough cone, like the carbon of an arc lamp, while the platinum on the blade is eaten out cup-shaped similar to the second carbon. It would therefore appear that the metal was volatilized at one point and deposited on the other. The probable reason for this may be the use of an inferior platinum alloy, which has a comparatively low melting point, and is, therefore, more readily volatilized than pure platinum. Pure platinum does not stand the rapid knocking at the high speeds the trembler works at, and hence an alloy of platinum and iridium is largely used, and this probably follows the rule that the melting point of an alloy is lower than that of either of its constituents.

In many cases frequent adjustments of the trembler screw and trimming up of the points only cure the trouble for a time, and lead to the conclusion that there is also some inherent fault in the coil itself.

Extra Grounding Wire.

"My 6½ H.P. single-cylinder engine had always been a trouble to start, but once going would run well," writes a European motorist. "Complete rewiring, a new sparking plug, shifting the non-trembler coil nearer to the engine, and a general clean up of all the electrical fittings did not improve matters. Finally, a second ground wire was attached to the coil, ending at the blade of the contact breaker, and the engine now starts without hesitation. The question now is, Why was the first ground wire ineffective? It was a good wire from the coil to a bolt holding the engine to the frame of the car. The platinum-tipped screw is insulated, but the trembler blade is attached to a metal segment moving about the half-time shaft as usual. The only explanation I can give is that the thick oil from the crankcase proved an insulator, a film of this lying around the half-time shaft between it and the metal segment. While the engine was at rest this oil more or less set hard. Upon trying to start the engine, it was necessary, by a long period of starting handle exercise, to wear through this film of oil until the one metal surface of the half-time shaft rubbed against the other metal surface of the segment. This has now been saved by giving an alternative path direct from the blade to the coil."

The Trembler Coil.

Many motorists would be glad to have an explanation of the reason why a trembler coil is necessary with a wipe contact, and the difference between an ordinary coil without and the coil with a trembler. To summarize the reason, it is necessary to break up the primary circuit of the coil rapidly— that is, the current which flows from the source of electrical energy through the coil. This interrupts the lines of the magnetic field, and intensifies the power of the induced current, for the quicker the make and break at the trembler, the more effective is the spark, or, rather, the shower of sparks, at the plug, such a shower being much hotter than those of lesser density produced by a slow vibrating trembler. That is why, as a rule, the magnetic trembler is much more effective than the mechanical trembler, for the latter cannot work up to the speed of the former.

A Cause of Misfiring.

An engine, after being in use for some time, will often misfire, and the cause be difficult to ascertain. Frequently the trouble is one which does not occur with a new engine. If the contact breaker is of the make-and-break type, the bearing of the contact maker may have worn and the current find difficulty in reaching "earth." To obviate this, a stranded wire, which need not be insulated, may be led from some part of the motor to the screw which holds the trembler blade in place. Then the current travels directly from the trembler to the earth without having to go through any sliding joints. etc. In a wipe contact maker, it will be often found that the brass is worn flush, or lower than the insulation, the result being that the wiping blade jumps the brass at high speeds. The insulation can easily be removed near the brass, which will cause it to operate as before. Contact breakers of this type should always be oiled, as it prevents the wiping blade carrying the insulation on to the brass, which it frequently does owing to its wiping action, the result being misfiring at high speeds. Again, misfiring such as this is often due to the wiping blade having lost its stiffness, and not bearing sufficiently hard upon the contact breaker cam, which may have worn as well.

Misfiring Through Defective Insulation.

Much annoyance is often caused to the user of a motor car by occasional misfiring which he finds a difficulty in locating. We cite a case of this kind, where for a time the engine would run perfectly, and then most unaccountably commence to misfire. This alternated with periods of regular running and irregular running. A thorough examination revealed no defects—the battery was fully charged, coil worked perfectly, and the contact maker made good contact—but still, as stated above, trouble was experienced repeatedly. It was noticed, however, on turning down the front of the coil box, that the high-tension wire of one of the coils was brought very close to the low-tension wire of the other, and the owner found that, instead of this high-tension wire being thoroughly protected by its insulation, at intervals a spark would leap through the insulation to the low-tension terminal before mentioned, and a misfire thus be caused in one of the cylinders. Removing the wire to a distance of about half an inch from the other terminal immediately corrected the fault, and no further trouble occurred.

In such cases of misfiring it is always advisable to inspect all wires which touch a metallic part of the frame, as, owing to the vibration of the engine, the cause may be at these points.

Marks on French Induction Coil Terminals.

The marks upon induction coils of French manufacture are not understood by a great many users, and therefore an explanation of them may be of interest. There are usually three terminal screws upon the coil for a single-cylinder engine, and these are marked P, M and B; P and M usually being at the side of the coil at the top, and B either directly at the bottom or lower down on the side. The terminal marked P is connected to the positive terminal of the storage battery by an insulated wire into which the contact breaker is interposed. There are, of course, many variations in the wiring of the connections to the coil, and the one given herewith is only one of them. The terminal M, which, by the way, on some French coils is also marked V, is connected to some metallic part of the frame on the car, or to the engine itself, and forms an earth or ground return to the negative terminal of the battery, a further piece of wire being connected from this terminal to the framework to complete the circuit.

The terminal B is the one to which the high-tension wire connecting up to the plug should be attached.

Numbering the Coil.

On three and four-cylinder engines fitted with trembler coils, it is always well to take an early opportunity of verifying the trembler for each cylinder. For instance, assume that a four-cylinder engine is missing on one or two cylinders. The bad cylinder is ascertained by holding down the trembler blades and making the engine run on one cylinder at a time. When we come to the bad one, the engine stops unless the other blades are quickly released. Despite this we find that all four tremblers are buzzing merrily in turn, and apparently there is nothing wrong with them. It is therefore natural to assume that the plug is at fault, and the question at once arises, which plug? There is nothing for it but to unscrew them one by one, and to turn the engine round to see if the one under examination is sparking out of the cylinder. In the usual course of things it will be the last plug one takes out which is found to be foul or otherwise at fault. On the other hand, if we had known which buzzer belonged to each plug, we could have gone straight to the foul plug and have cleaned it up or put in a new one without loss of time. However, as we have got our four plugs out, we might just as well count them off from the front. Call the one nearest the radiator No. 1. Turn the engine round slowly, and when you see the plug on No. 1 sparking, go round to the trembler coil and see which of the four tremblers is buzzing. Then pencil on the trembler case opposite to this trembler No. 1. Continue the operation until you have numbered the coil for all four cylinders. Of course, we know the engine does not fire 1, 2, 3, 4 backward, but that does not matter; we only suggest numbering the coil so that it can at once be seen which trembler and which plug are connected, so that in future, whenever there is a cylinder not firing, we can safely assume, when we have played the usual four-finger exercise on the coil, that the plug on a certain cylinder is wrong. We no longer need blunder through all four.

General Troubles with Coils.

Looseness of platinum screws in the bridges.—Whether these are bound with a lock-nut or not, they offer resistance to the primary current.

Armature rubbing against the guide screw.—This restricts the speed of the armature. Allowance is made for this in some armatures by making the hole in the armature through which the binding screw passes slightly larger than the screw stem.

Shorting of the secondary current to the coil support angles or ears, due to the screws which hold the angles to the case being too long and projecting inside.

In the case of four-cylinder coils (not waxed in entirely) the ebonite top breaks away, owing to the screws which are passed through the case into the ebonite top chipping out pieces of ebonite, and so losing their hold of the top. This is very often caused by the windings getting loose, and may be caused by a jar to the coil.

Breaking of the primary wire between the communication screws on the ebonite top and the terminals, and between the communication screw and the bobbin itself.

Breaking of the flex wire between the bobbin (high tension) and the terminal.

Internal switch troubles, due to wax entering the switch and greasing the metal contacts.

Buttons on the armatures (which draw down the platinum blade) getting loose and causing erratic striking of the platinum blade. In the case of some distributer coils this may be the cause of knocking in the engine.

Button of the armature shorting on the platinum screw.

Stiffness of the distributer armatures—in those cases where it is of springy material and has no spring underneath to help its return movement. This stiffness causes misfiring at high engine speeds.

Many owners have had cases of coils bubbling the wax out. This has been in most cases where ordinary coils have been used for distributer purposes. They have noticed also that with certain distributer coils with the bobbins waxed separately all the insulation off the bobbin melted down to the bottom of the box. This probably was due to the coil being placed inside the bonnet near the engine.

In large heavy two, three and four-cylinder coils the wooden cases have split, especially where angles are screwed to the wood. As an opposite example to this, the case is cited of a four-cylinder coil where the wood of the case and front flap was half an inch thick. It looked a very substantial coil, and would probably stand rough usage. The top had hinges, and front flap hinges were also much stronger than usual. Where flaps in covered-in coils are used the hinges are often loose. The same applies to the top lids. This is accounted for by the thin wood used in the case construction, which necessitates small screws being used to screw on the hinges.

Commutator Short Circuiting.

Good as rolling contact commutators undoubtedly are, yet nevertheless trouble may arise from them and within them which the automobilist may be long in diagnosing if he has not been informed that its happening is within the bounds of possibility. After considerable use, the friction of the roller over the metal contacts has the effect of wearing off small particles of the metal and gradually laying these on the fiber ring in the form of an embedded train, which will ultimately connect one contact with the other, so that the current will short all round, and whether two or four cylinders are served, current will pass to all the sparking plugs at once. If this has happened, the only thing to do is to detach the fiber ring and scrape the inlaid metal from its surface. These remarks apply with even more force to the commutators made on similar lines, but which have a rubbing in lieu of a rolling contact.

A Mysterious Squeaking Noise.

Sometimes a motor will develop a mysterious squeak when running, and this often takes a good deal of locating. Many motors are fitted with the wipe type of contact maker, and it is well to look to the wiper blade and the disk on which it rubs for the source of the squeak. If the disk is allowed to get dry, a most distressing noise is caused by the rubbing of the steel wiper piece on the fiber of the disk or by the bearing of the roller on the wiper arm when the latter is rotated. A spot or two of ordinary lubricating oil will effectually cure the trouble.

Adjusting Commutator Chain Drive.

If a chain-driven commutator upon the dashboard is used, it should be remembered that a great deal of difficulty will be experienced in getting the chain correctly replaced if it is taken off for any purpose. Of course, where a spur gear is employed to drive the commutator, it is perfectly easy to mark one tooth and the bottom of the two opposite teeth into

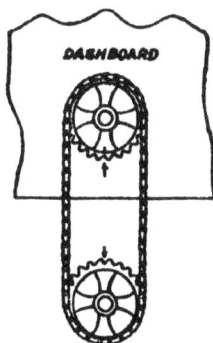

which it engages, thus insuring correct timing; but with a chain drive it is impossible simply to mark two teeth alone. This may be done to a certain extent with satisfaction, however, by marking the rim of the wheel on the center line; it then becomes a matter of the eye in replacing the chain, and also one of memory to insure the marks being in correct position, that is, both marks should be at the bottom or top of the wheel, as originally placed when they were indicated. One writer suggests pointers being attached to convenient parts, the marks on the chain wheel being brought opposite to these. In any case, it will, of course, be necessary to ascertain roughly the relative position of the crankshaft to the camshaft. Unless this is done, it is quite possible to get the setting incorrect, as while one wheel may be in position correctly, the other may be a revolution before or behind it.

Twisting Temporary Connections.

In making a temporary electrical connection, the stranded wires should be twisted up as solid as possible, and the loop formed by turning the wire from left to right. When so made, the loop closes in under the twisting action of the screw when tightening up the connection. If the loop be made in the opposite direction, this same action spreads the wire, and a bad connection results.

Contact Breaker and Commutator.

The terms contact breaker and commutator are at present being very loosely used in connection with automobiling, simply on account of their functions not being definitely and clearly understood. To all intents and purposes, both serve the same purpose, which is that of an automatic switch completing the circuit of an electric current at a given time. The contact breaker, whether it be of the spring blade or of the wiping contact type, is used in connection with single-cylinder engines only, while the commutator is used on multi-cylinder engines, though its type and design may be precisely similar to that of the wiping contact breaker on the single-cylinder engine. The very word "commutator" should be sufficiently expressive to prevent this error, as its meaning clearly shows that its mission is to "commute," or to exchange, the current from one path to another—that is, of course, from one cylinder to another. It could only be a contact breaker when each cylinder was supplied with a separate source of electrical energy, and with a separate coil, though when the common source of supply is from a single storage battery, notwithstanding that it traverses a separate coil for each cylinder, it then becomes a commutator.

Multi-Cylinder Ignition Timing.

There are still a few makers of four-cylinder engines who adhere to make-and-break ignition contacts as their standard, and when perfectly adjusted and tuned up, this ignition is quite as satisfactory as the trembler coil and wipe contact; but when the slightest derangement occurs, the trouble is difficult to locate, and often inexplicable. Presupposing that all the platinum contacts are in good condition, and that each cylinder is firing in its turn, it is yet quite possible that any-thing but the best results are being obtained. The defect arises solely from faulty ignition timing, due to the fact that the points of the platinum-tipped screws and blades are not all equally adjusted. Thus, if we suppose our four tremblers to be adjusted with No. 1 set of points 1 millimeter apart, No. 2 set 1.5 mm., No. 3 set 1.4 mm., and No. 4 set 1.2 mm., the cam having a 3 mm. eccentricity, each and every trembler will give a spark at its full power; but if we suppose that trembler No. 1 is firing accurately, No. 2 is firing late, No. 3 late also but earlier than No. 2, No. 4 earlier than either No. 2 or No. 3 but later than the correct No. 1, the terms late or early being, of course, relative to the position of the piston. Thus, in each cylinder the mixture is being ignited at a different period, with the result that, if No. 1 is being fired to its best ad-vantage, the other three cylinders are not igniting efficiently, the balance is gone, and considerable power is being lost. Be-yond this, where the firing is late, the combustion is not com-pleted until after the exhaust valves have opened; the burn-ing charge passes out in the form of a flash, extremely detri-mental to the exhaust valve heads, and tending to overheat the engine. In order that the best power may be obtained, each cylinder must explode at relatively the same point, and, therefore, when adjusting the make-and-break mechanism, great care should be taken to see that exactly the same distance separates the contact points.

On Preignition.

If preignition occurs, the engine should at once be stopped and examination be made, as it is the chief cause of bent connecting rods and broken crankshafts. The chief cause of preignition is failure of water circulation. If this should be suspected, it can be proved easily whether it is at fault by placing the hand on the radiators and water-jacket. If the former are almost cold, while the cylinder jacket is exceedingly hot, it will at once be understood that the pump is not working, or that there is an air-lock in the radiator piping. The best way to deal with the failure in either of these cases is to disconnect the outlet pipe from the pump and run the engine. If the water is not discharged, it is obvious that the pump is not working, while, on the other hand, if the water tank is filled up while the engine is running, the air-lock will probably be removed. Intermittent preignition is rather more dangerous than persistent preignition, because the latter pulls the engine up, while the former, coming at rare intervals, is far more likely to do damage owing to the driver neglecting to take precautions. A frequent cause of intermittent preignition is a short circuit in the contact breaker wire. This may be due either to the insulation of this wire becoming chafed and short circuiting to "ground," or to an errant strand at the contact breaker terminal, which frequently touches some part of the motor and "shorts."

A third cause puzzled an experienced motorist for a long time. The symptoms were of persistent preignition. The pump was suspected and overhauled. The water was emptied away and the tank refilled with cold, but before the car had traveled a hundred yards the owner was forced to stop the engine again, when, of course, he knew it was not the water circulation, but something inside the cylinder. On removing the valve caps, he found in the cylinder pieces of porcelain which had broken off the inside of a porcelain sparking plug. The central wire was held in place, and the plug was firing quite normally. These pieces of porcelain naturally got excessively hot, and caused the preignition. The owner had difficulty in getting rid of the pieces, but when removed preignition ceased.

Experiences: The Value of Diagnosis.

The foregoing hints are the result of an interesting personal experience, for faulty ignition resulted in a 16 H.P. car doing but sixteen miles an hour on the level with its levers in the

45 m.p.h. position, and reduced to a miserable crawl up anything like rising ground. The symptoms at first were precisely similar to those produced by the butterfly throttle valve having become loose on its spindle, but this was after a time proved to be an incorrect diagnosis. Next the governor was attacked, but, being found in order, the operator looked to the carbureter, rather expecting to find the gasolene supply choked to a slight extent, but everything was found clear. Between these investigations the car was run for a few miles, so that a chart of times and distances would have presented the appearance of the toothed edge of a saw. However, the operator tackled the ignition, and soon found a somewhat considerable blowing at the ignition plates (this being tested by pouring a little oil around the spindles of the tweakers) and considerable maladjustment. So much for the value of theory and diagnosis. We do not deprecate the system of observing symptoms and following them out to the end, for in a high percentage of cases the correct trouble is found, but, as we have shown, on occasions one is apt to be led far away from the actual ill. The ignition in the above case, by the way, was low-tension magneto.

Insecure Terminals.

Cars are sometimes sent out with stranded connecting wires just twisted round all the terminals, and there held by the screwing up of the terminal screw. We would strongly advise any automobilist who finds his new car wired in this careless and shiftless fashion to get proper terminals soldered on without delay. It will save both time and temper in the long run. Moreover, from frequent bending round the terminals, the stranded wire breaks, and one often gets nasty, painful pricks in the fingers therefrom, which smart and are sore for some time. There can be no sort of excuse for sending out cars wired up in the slipshod way we have referred to, and the purchaser of a car should see that it is put right.

Varnish for Electric Terminals.

Electric terminals which happen to be in such a position as to be subjected to water or mud accumulating upon them can be effectually prevented from possible short circuits by painting them with a varnish composed of ordinary red sealing-wax dissolved in a little gasolene. This varnish is made by putting into a small bottle a quantity of small pieces of sealing-wax, covering the latter with spirit and occasionally shaking it. If the varnish should prove too thin, add a little more wax or leave the cork out of the bottle until some of the spirit has evaporated. If it is too thick, add sufficient spirit to bring it down to the required consistency. In order to prevent the varnish retaining the brittleness of the sealing-wax, a little linseed oil should be added. For those who do not care to go to this trouble, a little melted paraffin-wax can be used for the same purpose. The ordinary wax candle contains paraffin-wax of sufficient quality to do this. Either of these methods has been found as satisfactory as binding with insulating tape.

Making Electric Connections.

A sketch of an excellent method of making electric connections with the wire itself is given herewith. The insula-

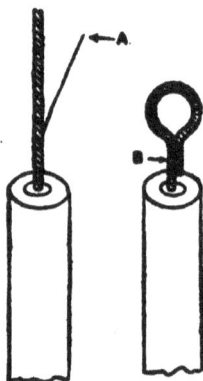

tion must be cut round at a convenient distance from the end, 1¼ inches to 1½ inches usually being the extreme amount required to make a connection. The stranded wires should be twisted tightly together; one or two of the wires, according to the thickness of the strands of which the cable is composed, are taken apart, as shown by A, the cable then being retwisted. The wire should then be formed into a loop round a piece of metal or the terminal itself to a nice easy fit. The end of the wire after forming the loop should lie parallel to the wire at the beginning of the loop. The stranded wires which have been taken apart are then used to bind the end of the loop to the main body of the cable, the whole being soldered together with soft solder, which will flow easily without having to use a great deal of heat. Particular care should be taken to use resin, instead of hydrochloric acid reduced by dissolving zinc in it, or one of the many acid soldering fluids sold. The objection to using such fluids is that they set up corrosion and a chemical action at the joint, offering a high resistance to the current, and there is no doubt that the same cause is responsible for the ignition delays which some motorists experience with their cars.

Broken Plugs.

Sparking plugs with loose and leaky centers are by no means uncommon; but, treating the matter generally, very few people try to discern and remedy the cause of the trouble. More often than not the source of breakage can be traced to the festoons of heavily-insulated wire pendant from the plug terminals, or, where a neater and collective arrangement is employed, to the tighter wires from the overhead stay to the plugs which transmit the vibration to a considerable extent, resulting in a breakage. There are two methods by which this transmitted vibration can be obviated entirely, and the life of the plug increased considerably. The one is to solder a fine coil of flexible wire to the end of the high-tension cable, support the cable firmly, and connect up the remaining end of the coil to the plug terminal. In this manner the weight of the cable is taken completely from the plug, and the fine coil is quite incapable of transmitting the vibration.

A similar arrangement—one which performs the twofold functions of spark gap and non-vibrating connection—is now fitted to a number of engines. From each high-tension terminal to the plug terminal is fitted a light brass or silver chain, down which the current runs to the plug, having a minute spark gap, as a rule, between each link; but all uncovered sparking gaps are dangerous.

Spark Plug Troubles.

If the porcelain body of a sparking plug allows a loss of compression at the packing gland, it is often only necessary slightly to tighten up the hexagonal top of the circular portion of the gland. After doing so, the plug wire should be inspected to make sure that any slight rotation of the porcelain does not affect the adjustment of the two points, otherwise some misfiring or entire failure to fire the charge may result. Another frequent cause of maladjustment of these points when the plug is new is the screwing of the plug into the cylinder. When the wire attached to the metal body of the plug is hammered into position the thread is usually burred slightly. This is restored to position when the plug is screwed into the cylinder, and the wire is slightly moved in consequence. When new a plug should be filed at the thread by means of a triangular or fine half-round file, to remove the bur. The plug should be screwed home, and then removed and examined to see that the position of the wires is not varied, after which the plug can be again screwed into the cylinder, with the certainty that it will work correctly.

Warped Spark Plug Porcelain.

An Eastern motorist experienced a rather uncommon failure, accompanied by misfiring and a peculiar blowing noise in one cylinder. As the valves had recently been ground-in in all the cylinders, and no compression cocks were used, the trouble could not be located until kerosene was squirted around the valve caps and over the sparking plug. Turning the starting handle round, it was found that gas blew past the porcelain of the plug of the misfiring cylinder, though from what one could see the plug was in proper condition. On removing the plug and taking it to pieces, the porcelain was found to have warped considerably. There is no doubt that the porcelain was not true in the first place, and from some unknown cause the packing which secured it had loosened sufficiently to allow an escape of gas past it, and so caused the trouble.

A Crack in the Porcelain.

When misfiring takes place, one usually in the first instance examines the sparking plug, which is supposed to be the offender, for deposits of sooty matter or lubricating oil. In a number of cases it will be found that when the soot or oil on the porcelain has been washed off with a little gasolene, and the sparking points cleaned with fine emery or glass-paper, a very good spark is seen between the points when the metallic body of the plug is laid on the cylinder and the necessary contacts made. Yet on replacing the plug it is found that the misfiring in this particular cylinder is just as bad as ever. This is a most deceptive and annoying trouble, which will often be caused by a crack in the porcelain, either close to the wire terminal and almost imperceptible, or it may be somewhere inside the body of the plug, and therefore cannot be seen. A good spark is produced in air, but under the compression at working conditions the spark passes from the center wire through the crack in the porcelain to the metallic body of the plug, as this offers a relatively easier path than that between the points and through the compressed mixture. If the plug is held with the metal body in one hand and the porcelain in the other, and a twisting action backward and forward is applied while the plug is held close to the ear, a slight grating sound will be heard if the break is inside the body of the plug, which should be at once discarded in favor of a sound one. A new porcelain may be fitted to the defective plug if desired. Great care should be exercised in putting in or taking out plugs from the cylinder, as there is every chance that the porcelain may receive a slight tap with a spanner and be broken, porcelain being extremely brittle. This particularly applies in cases where plugs are placed in deep recesses and a box or tube spanner is required for insertion or removal.

MODERN BATTERY SYSTEMS.

By the "Modern Battery System" is meant that type of ignition which comprises a single non-vibrating spark coil, a high tension distributer and a circuit breaker as described in Part II and illustrated diagramatically in Fig. 2 on page 44. Since only a single coil is used for any number of cylinders, the high tension distributer is necessary for distributing the spark to the various cylinders in the proper firing order. To understand the principles of repair mentioned in this section the reader should first consult the text in Part II and follow the circuit diagram Fig. 2-3 carefully.

As the circuit breakers of the newer systems are almost invariably of the platinum contact point variety with a swinging arm and cam, the methods of repair are different than in the older system in which a wiping roller contact is used. The high tension distributer also introduces new problems not discussed under the older system. As with all high tension apparatus, the breaker and distributer must be thoroughly protected against the entrance of moisture, oil or dust. Attention must also be paid to the insulation and the support of the wiring. Grounded contacts or connections must be frequently examined at the point of connection with the frame. Any looseness in the shafts, contact arms, control rods or any tendency for the timer to "wabble" on the shaft should be immediately corrected.

In cases of misfiring open the primary breaker box and examine the platinum contact points. If they appear dirty or oily clean them carefully with a lintless rag, taking care that no threads remain in the box after cleaning. If they now appear burnt, pitted, or have irregularities on their contact faces, carefully file the opposing faces smooth and square with a fine file. The finished surfaces should have a flat, even bearing, for should they come together on an edge or corner there will be heat generated, the resistance will be high, and pitting will again proceed rapidly. After filing clean out the metal filings and dust by means of a fine brush.

Now examine the levers, arms and cams for looseness or end play. A small amount of lost motion will cause error in the timing and will grow worse rapidly if not immediately repaired. Remove the levers from their shafts, clean thoroughly and apply only a single drop of thin oil on the shaft by means of a toothpick. Avoid overoiling or spattering oil on the contact surfaces. A single oiling of this nature should be good for 500 miles. Examine the springs that press the followers or rollers on the revolving cam. See that they hold the follower firmly against the cam face, for these springs often weaken and allow the follower to jump over the "high places" on the cam and destroy the timing. When the engine misses at high and runs well at low speeds it would be well to examine the spring tension.

Looseness in the timing lever connections that run from the steering column to the timer will cause uneven running for the reason that the lost motion will cause the spark to be intermittently advanced and retarded. A defective dash switch will also intermittently open and close the circuit and cause misfiring.

After filing down the contact points make sure that they are brought back to their proper relative position, that is, see that the contact point in the adjusting screw is away from the arm contact by the original clearance when the breaker is fully open. If the full open clearance between the points is not kept constant, the timing will be changed or misfiring will result through imperfect contact. Usually the distance between the points when fully open is about 1-64 inch or equal to the thickness of a heavy visiting card.

If the engine still runs poorly it is evident that the timing is wrong (providing the carburetor adjustment is all right). As an example of proving the timing we will give the instructions issued for the Atwater-Kent breaker, a proof that will apply to the majority of this class of breakers.

First, fully retard the spark lever and uncover the fly wheel so that the timing marks on the wheel and the pointer can be readily seen. Remove the cable from the spark plug in cylinder No. 1 and have the end of the cable held about 1/8 inch from the cylinder casting. Now turn fly wheel so that the pointer is near the mark "No. 1—T. C." and slowly rock the wheel back and forth. A spark should appear between the end of the cable and the cylinder casting at the moment that the mark on the wheel passes the pointer, for at this point the piston is at the upper end of the compression stroke. If the spark occurs any distance on either side of this mark, the timing is out and must be corrected. If the timing were correct before the contacts were filed, the chances are that the contact in the end of the adjusting screw is too far away from the contact on the swinging arm or contact spring.

Turn the engine over until the contacts are wide open, and remove one of the washers under the head of the contact screw. Turn the screw so that the screw contact point moves forward toward the arm contact point and until the two are separated by approximately 1-64 inch. Try the timing again. Should the timing still be out, examine the cam, shaft and fastenings to see if anything has slipped or worn out of place. Examine the points or teeth on the cam or the cam follower and note whether the engaging surfaces appear worn.

When finally adjusted, the head of the adjusting screw should be tight against the washers so that it will not work loose. In the Atwater-Kent system it should be remembered that the spark occurs at the instant the rider snaps back and strikes the contact spring. See page 47, Fig. 3. For similar adjustments on the Delco as installed in the Cadillac, see page 50, Fig. 8, and the text starting on page 52.

Before leaving the circuit breaker see that all of the wire connections are tight and that the wires are not broken through the rocking action of the timing lever.

Considerable trouble with the contact points can be avoided by occasionally reversing the direction of the current through the breaker. The current can be reversed by reversing the wires at the battery so that the wire formerly on the positive pole will be connected to the negative, or by reversing the wires at the timer.

The high tension distributer should be cleaned out periodically as the smallest deposits of dirt or oil are sufficient to short circuit the high tension current. Deposits of dirt are most common in distributers that employ a rubbing contact brush since the material rubbed off the brush collects in the casing.

Should the engine stop suddenly or refuse to start, carefully inspect the wiring for broken strands, abraded insulation, or loose connections, or test the batteries if dry cells are used. With a self-starting and lighting system, the lights are an index to the condition of the batteries and with proper ignition apparatus it is possible to fire the engine long after the storage cells fail to provide sufficient current for the lights, or when the lamps burn red. Naturally, a powerful coil with a good voltage behind it will give a "fat" and hot spark, but this may easily be offset to a certain extent by high compression. The best practice is, then, to find by experiment the most efficient position of the points, and to have a permanent guide in the form of a gauge to set them up to. This gauge may be a piece of sheet metal, and once obtained should be religiously preserved.

Screwing in New Plugs.

Before screwing a new sparking plug into its place in the cylinder head, always file down to the thread level both in the groove and over the diameter the small hump which is usually raised by the fixing of the sparking wire to the metallic body of the plug; otherwise, although the points are correctly spaced before screwing the plug into its place, there is a tendency to alter their relative positions on screwing in, and thus to cause failure in working. Always make sure that the center wire is fixed properly, otherwise if it can be moved readily the probability is that there is a bad connection, and trouble with misfiring will follow. Before putting in a plug, if the screw thread is lubricated with graphite or blacklead, it is more easily screwed in or out.

Periodical Examination of Spark Plugs.

Because your engine starts up well first time round every day, runs well to the ear, and seems to pull all right, do not leave your sparking plugs unexamined from one month's end to the other. You will insure the extra touches of speed and power if you take these fittings out from time to time, say once a fortnight, and scrape all the hard carbon off them, cleaning them finally and nicely with a stiff toothbrush dipped in gasolene. Engines with high compression will be improved by this little attention.

Simple Test for Short Circuits and Broken Circuits.

When troubles occur in connection with ignition, there is usually no method other than minute examination adopted for investigating the condition of the wiring in the primary circuit. Short circuits to the frame owing to chafing or cutting of insulation, and total disconnection owing to hidden breakages in the wire itself, are not unknown, so a simple means of testing for these faults may be of value. When batteries run down rapidly while at rest, the blame is put upon the cells themselves, but in reality it may be due to defective insulation of the primary circuit.

The instruments required for the test are usually ready to hand, these being simply a storage battery and a reliable voltmeter of fair size. On a peculiarity of the former the value of the test depends, for it is found that when a battery is on open circuit it gives a higher reading than when a current is flowing. The following actual readings illustrate this: Current flowing in primary circuit, 0, 1, 2, 3 amperes; readings of voltmeter, 4.05, 3.97, 3.92, 3.87 volts respectively. These readings can be taken advantage of for testing purposes as indicated below.

Short Circuit to Frame.

To test for a short circuit to the frame proceed as follows: Connect a voltmeter across the terminals of the battery (ac-

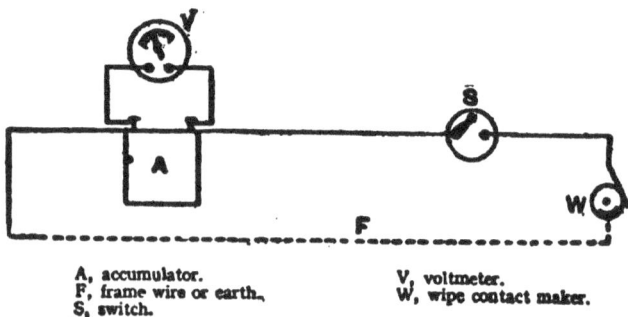

A, accumulator.
F, frame wire or earth.
S, switch.

V, voltmeter.
W, wipe contact maker.

cumulator); leave the switch open, and set the wipe or contact maker so that metallic contact is not made (see sketch). Take a reading of the voltmeter. Now close the switch, watching the voltmeter carefully while this is done. If the needle moves, showing a lower reading, it conclusively shows that a current is flowing through the switch; but since there is no connection at the wipe or contact maker (contact breaker, so called), this can only be due to a leakage.

Ignition by Batteries—Causes of Failure.

Below, under the four heads of Sources of Current, Induction Coil, Wiring, and Commutator will be found nearly every cause for ignition failure, partial or entire. Although the list is somewhat awe-inspiring as to length, every source of trouble enumerated can be avoided by care and cleanliness.

Sources of Current, Storage or Dry Batteries.—Loose or dirty terminal or terminal screw. Whole or partial discharge of storage or dry battery. Internal short circuit. External short circuit. Connection between plates or elements of the battery broken. Loss of electrolyte fluid. Perforated leaky casing.

Induction Coil.—Loose or dirty terminal or terminal screw. Wire separated from terminal at inner end. Broken or fused wire in coil (rare). Faulty adjustment of trembler. Pitted or unevenly burnt platinum contacts. Insufficient insulation of terminal from which high-tension plug wire leads.

Wiring.—Bared wire. Wire insulation destroyed by oil. Wire fractured within insulation. Loose wire attachment. Terminal unscrewed or dirty. Broken earth wire. Bad or dirty contact of earth wire. Mistake in remaking connections. Poor or faulty insulation of sparking plug.

Commutator, with Trembler or Blades.—Faulty adjustment of platinum contact screws. Too far apart—misses. Too near —premature ignition. Bad or burnt-out platinum contacts. Dirty or loose unsoldered platinum contacts. Faulty insulation of platinum-pointed screw. Greasy, split or loose ignition plate. Horizontal play. Broken trembler or contact blade.

Rotary Commutator.—Contact roller arm loose on spindle. Roller worn on circumference or center. Roller spindle worn. Contact ring worn in ridges, either fiber or metal. Too little lubrication. Too much lubrication with thick grease. Carbon deposit on contacts if contact maker runs dry.

Sparking Plug.—Fouling by oil, carbonization. Fracture of porcelain insulation. Loose porcelain. Shorting through mica insulation by penetration of grease. Sparking points too remote or too near. Damp, greasy or dirty porcelain.

HIGH TENSION MAGNETO TROUBLES

In general, the best advice that we can give the operator of an automobile or engine, is to leave any extended repairs of a high tension magneto to the service station of the maker, or to a firm making magneto repairs a specialty. In emergencies he may attempt minor repairs but the results are seldom satisfactory as it is an easy matter to ruin this delicate instrument through the neglect of the simplest precautions. Fortunately for the operator, the high tension magneto seldom gets out of order to such an extent that a general dissection is necessary, and hence it is well to go over every other part of the ignition thoroughly before setting the trouble on the magneto.

When the motor misfires or refuses to start, examine the plugs and wiring first, as they are the most frequent sources of trouble, not forgetting the carbureter. If the plugs are clean see that the spark points are not too far apart from the warping or burning action of the gas. The gap between the points should not exceed 1/64 inch, as high compressions reduce the jumping ability of the current. Due to the intense heat of the spark, small beads of metal often form on the spark points. These short circuit the spark gap and should be carefully filed away and the gap readjusted. Examine the porcelain of the plug for cracks and if oily or sooty clean with a tooth brush dipped in gasoline. Cracked plugs should be replaced with new.

The first point to examine on the magneto proper is the circuit breaker box and points. If the platinum points are dirty, or if there are evidences of oil in the breaker box, clean thoroughly with gasoline. Should the points be burnt so that they have an uneven bearing surface, or have small beads of metal attached at the point of contact, they should be very carefully filed to a flat, even bearing, so that they have a full contact over their entire surface. This operation should be performed with a very fine file, care being taken to have the contacting surfaces absolutely square with one another.

In the course of time the points wear down because of the incessant hammering and leave too great a gap between the stationary and moving contacts. This can be adjusted by means of the adjusting screw which carries the stationary point. This should be turned forward until the gap is not greater than 1/64 inch when the points are fully opened. A gauge for this adjustment is furnished with some makes of magnetos. After this adjustment is completed all lock nuts should be tightened up so as to prevent the screw from jarring back and out of position.

Irregular firing can only be caused by the improper adjustment of the breaker (when the magneto is at fault), and when the points have been cleaned and adjusted attention should be paid to the condition of the rocker lever, roller and other moving parts of the breaker mechanism. If the rocker lever has become jammed, thoroughly clean the bearings and axle by rubbing them slightly with fine emery paper, and then very slightly oil. In this connection it may be said that great care should be exercised in oiling a magneto as an excess of oil in the breaker box not only gums the mechanism but also prevents good electrical contact between the breaker points. If the breaker arm has been jammed make sure that it is not bent so that the contact points meet in an off center position.

If there are any current collection brushes in the breaker box see that they are thoroughly clean and that the spring presses the brush firmly against the corresponding contact point. A weak spring or dirty brush will interrupt the flow of current in the primary circuit of the magneto. Clean and tighten all wiring connection. Irregular firing is often caused by looseness in the levers and rods leading from the spark lever on the steering wheel up to the timing lever on the breaker box. Back lash or play in the driving gears or chain produce the same result. Thoroughly clean the high tension brush and holder that rests on the high tension collector ring, generally located at the rear end of armature. Clean ring and ring insulation with gasoline.

A dirty distributer or distributer brush is the cause of internal short circuits among the sectors to which the plug leads are attached. This produces irregular running among the cylinders as the deposits of dirt cause false connections between the different sectors and therefore intermittently connect the cylinders in the wrong order. Remove brush and thoroughly clean it and the distributer with a lintless rag dipped in gasoline. See that all high tension connections are free from moisture. Examine all wiring for abraded insulation, broken wires and loose connections, especially at a point near the circuit breaker where the rocking of the timing lever is likely to fracture the primary wires. See that the grounding wires are in good contact with the frame, and that the dash switch is in order.

Should irregular firing still persist, it is probable that the timing is at fault, or that the leads to the plugs are connected in the wrong order. This is very likely to be the trouble if the engine has just been overhauled or repaired. When replacing the magneto after a general overhaul, great care should be taken to mesh the teeth of the magneto gears at exactly the same place that they formerly occupied, for the difference of one tooth from the correct position will greatly distort the timing. The gear teeth should always be marked before removing the magneto so that there will be no trouble in replacing. This can be done by marking a tooth on one gear with a center punch, and then marking the two teeth (of the other gear) that lie on either side of the first tooth. See that the keyways in the gears and couplings are intact. Consult one of our wiring diagrams for connecting up the spark plug leads.

Should there be no marks on the gears, consult the directions for timing magnetos contained in Part XI, remembering that the circuit breaker contacts should just barely begin to open when the piston to be fixed is about ¼ inch from the end of the compression stroke. The order in which the leads are connected depends on the firing order of the motor, and the direction in which the distributer brush rotates. This subject is also covered in Part XI.

The magnets of a standard make of magneto do not weaken to any extent with age, even after several years of service. Should everything else be tried as outlined above, test for weak magnets by removing the magneto from the engine and running it at a speed of about 150 revolutions per minute. At this speed there should be an almost continuous spark discharge across the safety gap, if the magnets, armature, and condenser are in proper order. If the discharge is weak or intermittent it is evident that there is some internal disorder, and the chances are that the trouble lies in the magnets.

Owing to the many peculiarities of the permanent magnetic circuit we would not advise the amateur repairman to remove the magnets unless in the case of an emergency, but with care and attention to the following instructions it can be accomplished without harm, or at least permanent hurt to the machine. Go at the job gently, remembering that you are not handling a locomotive, but a machine that is more delicate than the average clock, and remember above all things that each part, screw and washer must be replaced in exactly the same place it originally occupied. Measurements on a magneto are not in 1/16ths but in ten-thousandths of an inch.

Due to the close running clearance of the armature in the tunnel, which is in the nature of 0.0015 inch, the alignment of the bearings and rotating parts must be as perfect as possible. Any error in this respect will cause the armature to strike the pole pieces or permanently bind in such a way that the entire magneto will be ruined if much force is applied to the driving shaft before the trouble is corrected. This small running clearance is necessary to preserve the strength of the magnetic field. Matters are further complicated by the pull of the magnets on the armature which tend to take up any slack in the bearings and bind the armature on the pole pieces.

When removed from the magneto the magnets will rapidly lose their strength if not carefully handled, owing principally to the great increase of resistance offered to the magnetic force by the removal of the iron circuit that formerly existed between the legs of the magnets. If not provided for, the magnetism will generally decrease at least 50 per cent in one hour after the removal. A complete iron circuit between the poles of the magnet must be provided if the magnets are to hold their strength. The spark obtained is directly proportional to the strength of the magnets and to the speed of the armature, a strong field causing the instrument to generate the current at comparatively low speeds.

When the magnets are removed from the instrument for remagnetizing, or for any other reason, never fail to place a soft iron bar or "keeper" across the two legs of the magnet on the instant that the magnet is removed, and keep the bar in this position as long as the magnets are off of the machine. When lying on the bench take care not to jar them unnecessarily by dropping them, or by striking with metallic parts. Do not pile the magnets one on the other, as the crossed and stray magnetic fields will tend to decrease the strength. Vibration, heat and external magnetic fields are detrimental to the permanance of the magnets. Better not remove them at all unless it is absolutely necessary.

Methods of remagnetizing are given in an article on page 227.

When replacing the magnets, after remagnetizing, be sure that the magnets are replaced in exactly the same place on the machine that they formerly occupied, and that the magnet legs of similar polarity are replaced on the same side of the armature. The best method is to mark the magnet poles and magneto frame so that there will be no possibility of mistake. If dissimilar poles are placed on the same side of the armature, the magnetic field will short circuit and decrease the effective flux. Carefully clean the poles at the point where they fit over the armature tunnel and screw them firmly into contact with the pole pieces. Any space between the poles and the pole pieces will decrease the spark and the life of the magnets. Do not place liners or shims at this point.

Difficult starting is more often due to improper timing than to weak magnets, and you may almost be sure that this is the trouble when the engine kicks on cranking with a retarded spark. In this case it will be found that the circuit breaker opens and causes the spark to occur long before the piston reaches the end of the compression stroke, with the result that the piston and crank is driven backwards against the normal direction of rotation. This of course means that the armature is too far advanced in regard to the piston. With the armature retarded too far it is impossible to develop the full speed of the engine even with a wide open throttle and excessive heat is developed at moderate speeds even with the spark fully advanced. One test of this condition is that the engine will not knock or hammer at moderate speeds with the spark fully advanced.

It must be remembered that both the primary and high tension current return from the coil, and from the spark plugs, to the magneto through the magneto base, and therefore there must be good electrical contact between the magneto frame and the frame of the engine (electrically called a "Ground"). The current can both return through the metallic contact of the magneto base and through the hold down bolts used in fastening the magneto. Be sure that coats of paint and paper liners do not interfere. If the grounded circuit is not complete the magneto will not generate.

With the engine running at about 150 revolutions per minute there will be almost a continuous discharge of sparks across the safety gap if there is a break or disconnection anywhere in the high tension circuit. This may be caused by a disconnected wire running to the spark plugs, or in transformer types may be caused by a break or disconnection in the high tension wire leading from the transformer to the distributer of the magneto. A broken brush in the distributer or broken connection from the collector ring brush will also cause a discharge across the safety gap.

High tension magnetos should be thoroughly protected against the entrance of moisture, or oil, for either are destructive to the insulation. Owing to the small armature clearance, the rust produced by moisture will soon weld the armature firmly to the pole pieces of the magnets, with the result that the shaft will be twisted off, or other serious injury will be caused. While modern magnetos are supposed to be waterproof, it is always better to be on the safe side and protect them with a leather or sheet metal cover. This should be of non-magnetic material. If iron is used it will tend to short circuit the magnetic field and reduce the spark.

When remagnetizing is tried, after the circuits and other parts have been found clear from trouble, the trouble is undoubtedly due to the derangement of the armature winding or to the condenser. If the magneto is of the true high tension type it is almost impossible for the amateur repair man to remedy the trouble, owing to the great length of fine wire on the secondary winding and to the careful construction of the condenser. It is far better in this case to return the magneto to the makers for replacement or repair. In the case of a transformer type where the armature winding consists of a few turns of heavy wire or copper strip, repairs can often be made successfully with the crudest equipment.

When removing the armature of any magneto take great care not to spring the shaft and do not lay it where dust or moisture will collect on the windings or core. A little rust or dirt will cause the armature to bind in the armature tunnel when replaced. A kink in the shaft or a bunged up bearing will cause the armature to strike the pole pieces or bind in the tunnel. Take care to place the bearings back into exactly their original positions and use the same bolts in the same holes. If there are dowel pins, drive them in gently before tightening the screws. In oiling the magneto it is not necessary to use more than two or three drops of oil per run of 500 miles on the main bearings. In some types there is no oil required in the breaker box, but in any event never use more than one drop at a time on the breaker lever and apply this carefully with a tooth pick so that it will not get on the contact points. Any light oil such as sewing machine oil may be used, but never cylinder oil. Gummy or carbonized oil on the breaker points will surely cause misfiring.

In cleaning the magneto, start at the breaker box, clean the distributor, and then clean the high tension collector ring and brushes. A soft rag, free from lint and moistened with gasoline, will remove all ordinary dirt and oil. Should the mageto become wet, dry slowly at a very moderate temperature (less than 150 degrees). Never in a hot oven.

In the majority of cases the trouble will be found in the plugs or in the wiring and these points should always be investigated first before tampering with the magneto. The first places to look at on the magneto are the breaker points and distributer.

SUMMARY OF TROUBLES AND REPAIR

Motor Stops Suddenly. Short circuit in low tension cable—Switch jarred into closed position—Moisture or dirt in magneto—High tension lead loose in wire running from coil to magneto distributer—Short circuit in high tension wire supports. **Test for Primary Leak.** Disconnect primary wire running from switch to circuit breaker and try to start engine. If engine starts without trouble there is a leak in the primary wire or trouble in the switch.

Misfiring. Dirty spark plugs—Cracked spark plugs—Spark gap of plugs too great—Disconnected plug wire—Faulty wire insulation—Loose wire connection in either primary or secondary—Poor mixture from carbureter—Dirty or burnt contact points in circuit breaker—Dirty distributer—Distributer brush makes poor contact—Breaker brush makes poor contact—Collector rings or collector brush dirty—Opening between breaker contacts worn or burnt so that the opening is too great—Breaker arm stuck or loose—Swinging grounds between wires and engine frame.

Starting Trouble. Spark plug gaps too great—Incorrect timing—Carbureter trouble—Starting motor runs at too low speed due to exhausted battery—Cylinders not primed—Circuit breaker stuck—Moisture—Closed switch—Short circuit in primary wires or defective switch—Disconnected plug.

Knocking. Spark advanced too far—Incorrect timing.

Instructions in regard to properly timing the magneto can be found in Part XI, page 227.

PART IX

ELECTRIC STARTING AND LIGHTING

General. When using electricity as a medium for "cranking" the gasoline motor it is possible to use the current also for ignition and lighting as well as for the electric horn and gear shift. The possibility of operating so many auxiliaries from the same source of power naturally makes the electric self-starting system by far the most popular. In many cases an independent magneto is used and sometimes in addition a third auxiliary, the dry cell system, is added to the ignition system making the car entirely independent of any one system for the ignition current.

Disregarding the ignition system for the time being, the self-starting and lighting system is composed of the following principal units:

(1) The generator for supplying the current for the cranking of the car and the lighting system.

(2) The motor for spinning the motor. (Sometimes the generator and motor functions are supplied by a single unit.)

(3) Storage battery for storing current for the motor and lights as well as for the horn and ignition.

. There are four ways in which a single unit may act as both generator and motor. (1) A single armature, field and commutator may give or receive current to or from the storage battery. (2) A unit with a single field and armature may be provided with two commutators and two independent windings on the armature, one winding being for the generator while the remaining winding and commutator is for motor service. (3) Two independent armatures, fields and commutators may be contained in the same frame, the armatures being mounted on the same shaft in tandem, they being electrically independent of one another during the starting and generating periods. (4) Instead of being in tandem the fields and armatures may be mounted in the same casing but one above the other. (Double deck.)

When types (1) and (2) are acting as generators they are generally driven by the engine through the timing gears. When operating as motors they drive the engine either through a gear toothed fly-wheel or by a silent chain to the crank-shaft. The driving pinion is so

160

arranged that it can be thrown in and out of mesh with the geared fly-wheel by the starting pedal, the gear being normally out of mesh when the engine is running under its own power.

Regulation of Generator Current. The faster the armature of a generator rotates, the higher will be the voltage, and the greater will be the current put through the storage battery. With a continually fluctuating speed due to the variations of the engine it is evident that some device must be provided that will limit the current sent through the storage cells and at the same time prevent the storage battery current from surging back through the generator when the generator falls below the voltage of the battery.

In general there are four ways of limiting the current. (1) By providing the generator with a governor so that it cannot exceed a certain speed. (2) By placing an automatically controlled resistance in the generator circuit that will keep the current steady at any ordinary speed of the generator. (3) Providing an automatic cut-out switch that will open the circuit when the current exceeds or falls below certain points. (4) By inherent regulation of a specially wound generator in which the windings oppose one another and diminish the output as the speed increases.

Double Unit System. There are several systems in which the generator and motor are entirely independent of one another and are mounted in different parts of the chassis. The motor is series wound while the generator is compound wound, the difference in winding being due to the fact that a series winding gives greater "torque" or pull while the compound winding tends to maintain a constant current.

The Cut Out. A cut out is an automatic switch which opens the generator circuit when the voltage of the generator falls below that of the battery so that the current from the battery will not be discharged back through the generator. This generally consists of an iron core on which a double winding is placed. One winding which is connected across the terminals of the generator consists of many turns of fine wire, while the other coil consists of a few turns of heavy wire connected in series with the circuit leading to the storage battery.

When the generator comes up to voltage, the fine wire coil magnetizes the bar so that the armature is drawn up causing the current to flow into the battery through the switch. The main current now flows through the heavy coil reinforcing the· magnetic effect of the first coil.

Should the generator now fall in speed so that its voltage is less than that of the battery, the current will be reversed in ,direction through the second coil which will therefore oppose the first coil, demagnetize the iron core and allow the switch to be opened by the tension of a spring connected to the armature.

Generating Speeds. All other conditions being constant, the speed of a dynamo or generator determines the voltage, the voltage increasing in almost direct proportion to the speed until the "saturation" point of the generator is reached. To obtain the desired voltage it is therefore necessary to have the generator run at a particular relation to the normal running speed of the motor. Gearing the generator at a high ratio allows the current to be developed at low engine speeds but also causes trouble at high speeds due to the tendency of developing excessive voltages and currents.

The exact ratio of the gearing between generator and engine depends of course on generating speed and the low limit speed of the engine. Again in the systems where the generator also carries the

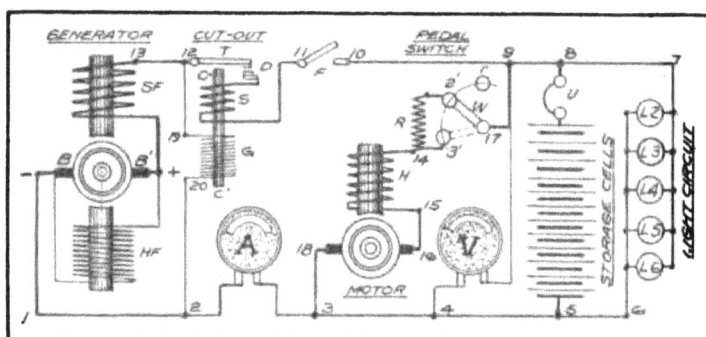

Conventional Starting and Lighting Circuit, in Which R is the Idling Resistance, D-T-C-S-G is the Automatic Cut-Out, V is the Voltmeter and A is the Ammeter.

NOTE! S is the series coil and G is the shunt coil of the cut-out, the circuit being opened by the switch T at the point D.

HF and SF are respectively the shunt and series windings of the generator. H is the single series winding of the motor.

primary circuit breaker and the high-tension distributer, the speed of the generator must bear a definite relation to the crank-shaft speed so that the breaker will cause the spark to occur in the cylinders at the proper time. Usually the generator gearing ratio runs from 1 to 1½ times the crank-shaft speed. This corresponds to a car speed of from 10 to 12 miles per hour.

Motor Speeds. To obtain the necessary torque to crank the motor, the speed of the motor, or the motor element of the motor-generator, must be much higher than the generator, and hence the gearing ratio must be higher in starting than in generating. This higher gear ratio gives the motor a greater leverage on the engine so that it can be spun under the most difficult conditions. In fact this leverage is so great that the starting motor is usually capable of moving the whole car at a low speed.

To obtain this reduction, and to save as many gears as possible it is the common practice to gear the motor, or the motor end of the motor-generator, to teeth cut in the periphery of the fly-wheel where a reduction is to be had of from 8 to 1, to 10-1. The motor pinion shaft is then provided with a second train of gears (usually encased in the motor housing) with an additional reduction of from 2 to 1, or 3 to 1. (See (R) in Fig. 1.)

Considering a normal engine speed of 1,000 revolutions per minute and a starting motor ratio of 30 to 1, it will be seen that the motor speed would reach a value of 30×1,000=30,000 revolutions per minute if it were left permanently engaged with the fly-wheel teeth. This speed of course would be entirely out of the question because of the mechanical stresses and the tremendous   so that the motor drive is always provided with some form of declutching mechanism that liberates the motor from the engine at a certain speed.

Usually it is in the form of a clutch which engages as long as the motor exerts a force on the engine, but when the engine exceeds the speed of the motor and reduces this force to zero or reverses it, the clutch will free the motor.

CONNECTING MOTOR AND GENERATOR TO ENGINE.

Connection between the generator and the motor to the engine depends to a great extent upon the arrangement of the engine and the other accessories, the three principal arrangements being as follows:

(1)   Geared connection to fly-wheel (already described).

(2)   Chain drive to crank shaft, or,

(3)   Through the magneto or pump shafts, and thence through the timing gears to the crank-shaft.

In addition to the above is the U. S. L. system in which the motor-generator is mounted directly on the crank-shaft in place of the usual fly-wheel. This is the simplest type of all since it dispenses with the usual gears, bearings, clutches and shafts of the other systems, and reduces the weight of the machine by an amount approximately equal to the weight of the fly-wheel.

When the drive is through the fly-wheel, with independent motor (M) and generator (G) as shown in Fig. 1, the motor end is cut in and out of service by throwing a pinion (A) in or out of mesh with the teeth cut in the circumference of the fly-wheel (F) by means of the starting foot pedal (P).

A switch (S) is opened or closed by the same movement of the pedal which opens or closes the circuit between the storage battery (B) and the motor. Depressing the pedal throws the pinion in mesh

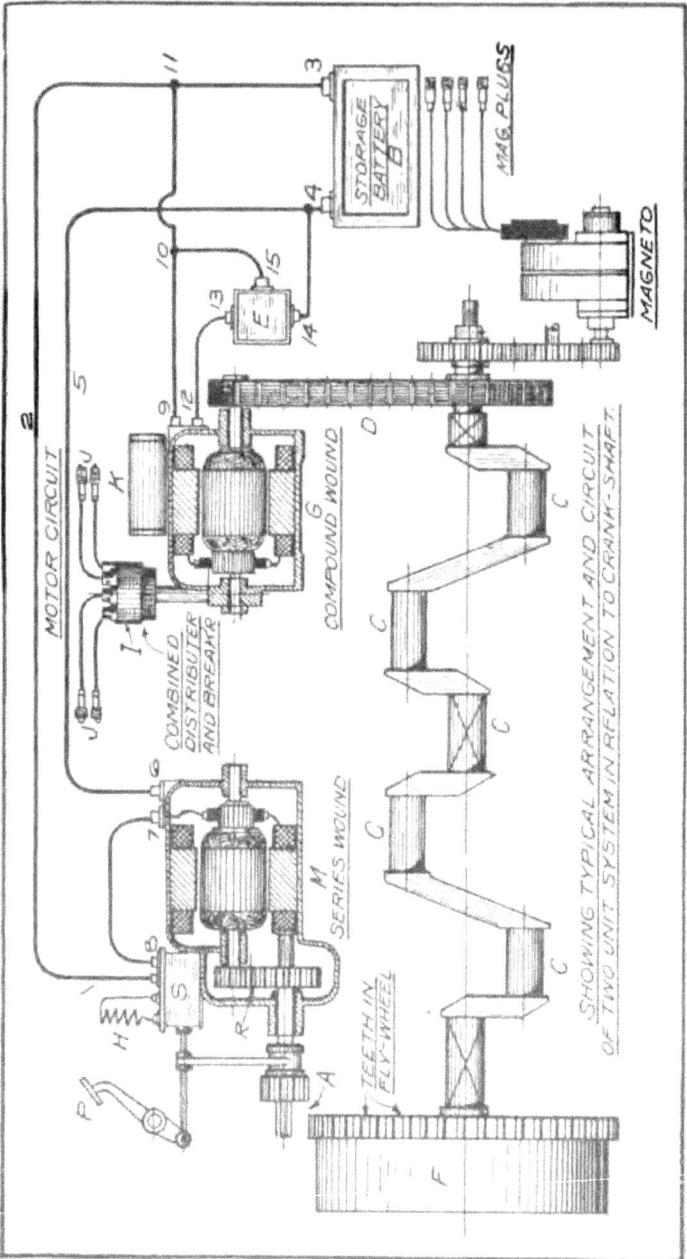

Fig. 1.—Two Unit System.

with the fly-wheel and closes the switch allowing the battery current to flow through the motor, thus turning the crank-shaft over and starting the motor. The second set of reduction gears is shown at (R). A resistance coil (H) is generally put in series by the switch which allows the motor to turn over very slowly until the gears are in mesh. When the pinion is forced clear across the face, the further movement of the switch short-circuits the resistance, allowing the full current to flow and the motor to build up its full speed.

The independent generator (G) is shown in driving relation to the crank-shaft (C), the drive being through the silent chain (D). The generator in this system is always connected to the crank-shaft no matter whether the engine is starting or running normally. A cut-out (E) is shown in series with the generator circuit (the purpose of the cut-out was described in an early part of this chapter). The distributer (I) is shown on generator feeding the spark plugs (J), the coil being at (K).

In Fig. 2 is shown the motor-generator arrangement in which the functions of motoring and generating are performed by a single unit (L). When starting, the pedal (P) meshes the pinion (A) with the fly-wheel gear teeth (K) as before described, the switch (S) performing the same way as in the two unit system. An extension of the armature shaft is driven through the gear train (I) when the unit is running as a generator.

Since two speeds are required for motoring and generating it is evident that some form of slip clutch must be provided as at (M) so that the armature (G) will be disconnected from the gears (I) when the motor is starting the engine and is running at a high speed. This clutch is usually of the ratchet type which will allow the armature shaft to run faster than or to run past the gear (N).

It will also be seen that with this type there are two independent commutators (D) and (E) for the two windings on the armature (G). A single pair of poles (H)-(H) serve for both the motor and generator windings.

Voltage and Battery Arrangement. In general there are three voltage arrangements at the present time, a straight system where lights, motor and battery operate at six volts; a straight twelve volt system; and a mixed system in which a double six volt battery supply current at twelve volts to the motor and at six volts for the lamps, horn and ignition system.

With the mixed system, the twelve volt leads are connected from the end terminals of the battery, while the six volt circuit is obtained by a third wire connected to the middle cell. A connection made between this third wire and any of the others gives six volts.

With twelve volts, either twelve volt lamps can be used or two six

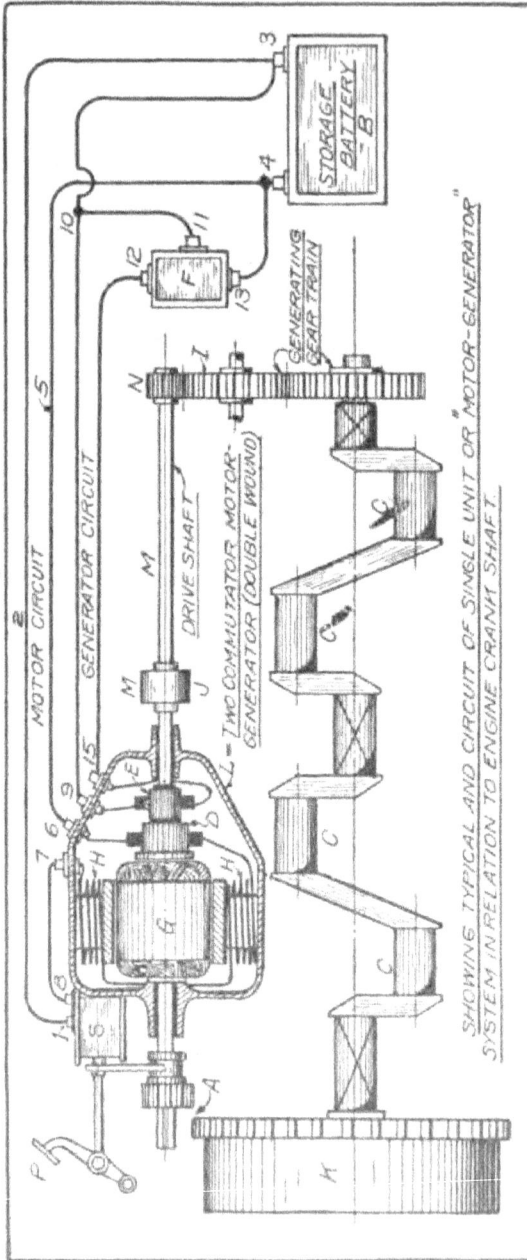

Fig. 2.—Single Unit System.

volt lamps connected in series across the twelve volt wires. The latter method is not the best as one lamp always burns hotter and dims the other. Unequal burning shortens the life of the lamps.

IGNITION LAYOUT.

It is possible to provide three absolutely independent ignition systems where the electric selfstarter is used. (1) Current from the storage cells. (2) High tension magneto. (3) An auxiliary dry battery. An independent system in which the starting and lighting storage cells are used exclusively is not desirable for the reason that the voltage is often much reduced owing to repeated demands for current in starting. The dry battery is by far the simplest auxiliary for the reason that a common circuit breaker, distributer, high tension leads, and spark plugs can be used both for the storage and dry cells without change. The use of a magneto involves a separate set of plugs and ignition wires as well as an additional train of driving gears.

The use of more than one set of plugs is ordinarily objectionable, especially with eight and twelve cylinder motors owing to the difficulty met with in arranging the wiring on the tops of the cylinders.

ELECTRIC GENERATORS AND MOTORS.

When a current conducting object is moved across a magnetic field, in a direction perpendicular to the magnetic lines of force, an electric current is generated in the conductor which flows in a direction at right angles to the line of motion. It is upon this relation between a moving conductor and a magnetic field that the operation of the electric dynamo (generator) is based, the essential elements being a stationary magnetic field and a moving mass containing the current generating conductors commonly called the "armature." In practice the conductors are in the form of copper wire coils which are rotated between the poles of a powerful electromagnet.

The electrical pressure or voltage depends upon the relative velocity between the conductors and the field and also upon the intensity of the field. The current output is limited by the size of the armature conductors or more directly by their electrical resistance. From this it will be seen that, other things being equal, a generator will deliver equal voltages at lower speeds with an intense rather than with a weak magnetic field. The voltage also depends upon the number of armature conductors, the greater the number of conductors with an equal speed and magnetic flux, the greater will be the generated voltage. The current is independent of the number of active conductors.

Aside from a few minor details of construction, the dynamo and motor are identical, both having similar armatures, magnetic fields, brushes and commutators. Any direct current dynamo can be run as a motor, while almost any comparatively large motor can be run as a dynamo. In the smaller sizes of motors, however, the air gap, or distance between the field poles and the armature, is much greater in proportion than in the larger sizes, making it a difficult matter to generate current when the field current is supplied by the armature.

THE DYNAMO OR GENERATOR.

In Fig. 4 is shown a section through an elementary dynamo, the dissimilar magnetic poles being indicated by the letters N and S.

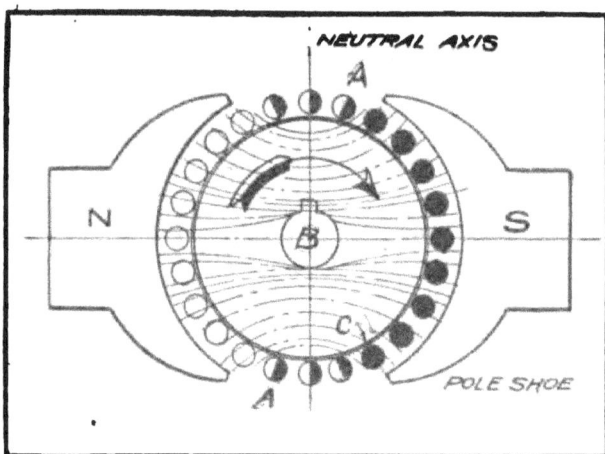

Fig. 4.—Elementary Dynamo.

The conductors, shown in cross-section, are shown by the small circles marked "A" mounted on the periphery of the armature core C, which in turn is mounted on the shaft B. The direction of armature rotation is shown by the arrow.

With a constant direction of armature rotation, the current induced in the armature conductors reverses its direction of flow every time that the conductor passes from one pole to the other, the current produced in the armature being alternating in character. If direct or continuous current is desired some device must be introduced into the external circuit to reverse these current waves into a constant direction. In the figure, the direction of flow in the conductors is indicated by the color of the small circles representing the armature conductors, the white circles at the left indicating a current that is flowing toward the observer while the black circles at the right indicate a current that is flowing back along the core.

It will be noted from the figure that the flow reverses from one pole to the next, the black conductors being under the left pole and white conductors under the right. The general direction of the magnetic flux is indicated by the curved dot and dash lines that extend from one pole to the other. As the three extreme upper and lower conductors do not extend through the magnetic field it is evident that they generate no current. The armature core C not only affords a support for the conductors but also increases the intensity of the flux through the conductors as the magnetic resistance of the steel core is less than the same length of air gap.

This core is always "laminated," that is, built up from circular discs of sheet steel. Before assembling the discs are painted or varnished so that adjacent discs do not come into electrical contact. Laminating prevents the generation of useless and harmful currents in the iron of the core, thus reducing the heating and power loss to a minimum. The copper conductors are of course thoroughly insulated from the iron of the core.

Since the direction of current flow in any one of the armature conductors is continually being reversed, due to the effect of the magnetic poles, it is necessary in all direct current machines to employ a form of rotary switch known as a "commutator" for the purpose of rectifying the direction of the reverse waves in the external circuit.

The commutator is a cylindrical drum built up of copper bars, the bars' length being parallel to the shaft of the generator. A connection from each armature coil leads to a single bar and each bar is thoroughly insulated from its neighbor by means of a thin strip of mica. Thus by going progressively around the commutator it is possible to make electrical contact with each of the coils independently.

Two conducting strips or "brushes" are arranged in insulating holders which make sliding contact with the outer surface of the commutator bars, these bars making contact at diametrically opposite points. As the armature revolves the bars successively make contact with the brushes, thus allowing the brushes to collect the current from each armature coil at the time when they are occupying a definite position in the magnetic field. Since a conductor gives only one direction of current flow in a single position in the magnetic field it is evident that the brushes always collect current of a single polarity.

With the type of electrical generator commonly known as a "magneto" the magnetic field is of the permanent type, that is, composed of permanently magnetized hardened steel bars. In the generators used for starting and lighting it is not advisable to use a permanent field, the common practice being to use a field produced by an electromagnet. With a permanent field the voltage fluctuates with the engine speed to such an extent that it is necessary to introduce an automatically operated resistance in order to maintain a constant charging current, a device that is generally avoided with an electromagnetic arrangement.

Fig. 5 shows the complete generator assembled in diagrammatic form, this particular machine being of the "Bipolar" or two pole type commonly used with lighting and starting generators. In this figure the armature conductors A are connected to the commutator bars C by the spiral lead wires shown. The brushes G and G' are diamet-

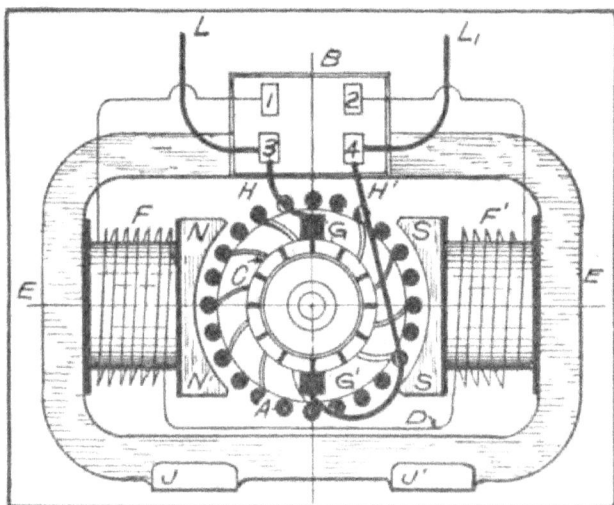

Fig. 5.—Assembled Dynamo or Motor in Diagrammatic Form.

rically opposite to one another on the vertical center line of the generator and the current from the brushes is led to the external circuit L-L' by the flexible leads H-H' that connect G-G' with the terminal blocks 3-4 on the distribution board B.

The magnetic pole pieces N-N' and S-S' are magnetized by the field coils F and F' which are wound with insulated copper wire in the same way as the primary coil of a spark coil. Current passing through the field coils causes a magnetic flux to pass through the armature windings and core in the same way as shown in Fig. 4. In some generators, the pole pieces are made of stampings or are laminated, while in other makes they are simply iron or steel castings. In any case the current required for the energization of the coils is obtained from the armature of the generator.

The iron or steel used for the magnetic circuit of the lighting generators is of a very soft grade so that the magnetization can be brought to a much higher point than would be possible with the hard steel used in magnetos. When the current ceases to flow in a field coil surrounding a soft iron or steel core the magnetism dies out almost instantly to a very small value.

A surrounding field frame E-E not only serves to support the armature and other parts but also acts as a return path for the flux on its way back from the armature core and pole pieces. In all types of lighting and starting generators the field frame in addition acts as a covering and protection for the interior wiring and the commutator against moisture and the oil thrown from the engine. The feet J-J' which hold the generator frame to the base of the automobile are in this case cast with the frame. When laminated fields are used, the laminations are held in a sort of sub-frame made of the cast iron, the laminations carrying the magnetic circuit while the sub-frame acts as a support for the various parts of the generator.

Armature current for the field coils enters at the terminal block 1, passes through the coil F, flows along the cross connection D, and then passes through the coil F' to the remaining terminal block 2. Connectors between the armature terminals 3 and 4 lead current to the field terminals 1 and 2. The exact interconnections between these terminals will be described later since they have a direct influence on the performance of the machine, dividing the generators into three different classes. Similar variations in the field connections of the motors also are used, the particular type used depending upon the use for which the motor is intended. The pole pieces N and S are always of opposite polarity as indicated by the letters, and the coils are therefore connected so that the current will flow in the same direction around the core.

In general there are four methods of placing the armature conductors on the core. Fig. 6 shows the methods commonly adopted. In "A" the conductors are laid directly on the surface of the core with a sheet of insulating material between the wires and the iron. The different coils are separated from each other by strips of fiber board which also act as supports while winding. After the winding is completed the wires are bound firmly to the core by means of transverse hands of brass binding wire. As this type must be wound by hand, it is difficult to repair, and is not proof against very high speeds.

At present the "Slotted" or "Toothed" type shown by Fig. B is most generally used, the slots affording a firm support for the conductors against the stresses set up by centrifugal force and then by driving effort. The imbedding of the conductors also is effective in decreasing the resistance to the magnetic field and therefore increases the efficiency. Circular slots are often used as shown by Fig. C, especially in the smaller machines, a small outer slot being used for the purpose of entering the formed coils. Fig. D shows full circular slots, the coils being entirely embedded in the iron of the core. This gives the greatest possible mechanical strength and fully protects the winding against abrasion. A construction of this nature is particularly desirable with starting motors or motor generators which under certain conditions may attain terrific speeds.

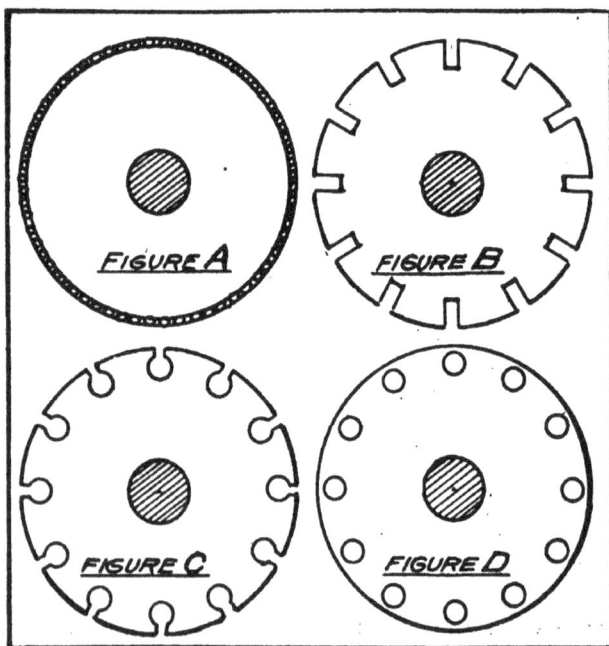

Fig. 6.—Types of Armature Coils.

As mentioned in a former paragraph, there are several ways in which the fields can be connected to the armature circuit. In Fig. 7 the three usual methods are shown diagramatically, the series winding, the shunt winding and the "short" compound. The particular winding adopted is determined by the voltage regulation of the generator or by the required torque and speed regulation of the starting motor.

In a series generator or motor the fields are connected in series with the armature circuit as shown by the upper diagram in Fig. 4. Current from the positive brush B passes through the lead C' to the field coil F, hence through C to the field coil F' and out to the external circuit at L'. After passing through the external circuit the current returns to the brush B', passes through the armature A and thence to the brush B, completing the circuit. In this type it will be seen, the entire current passes through the fields as well as through the

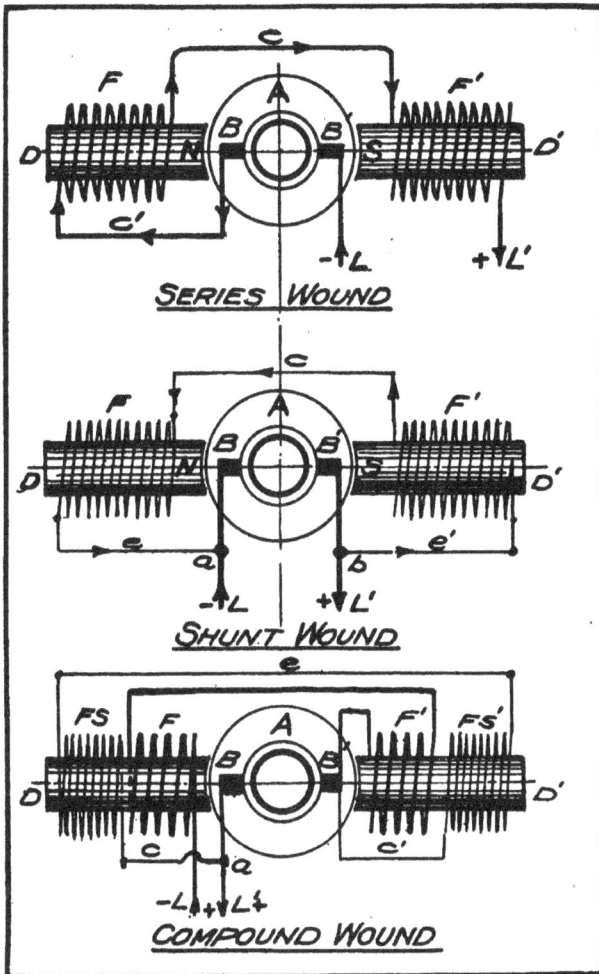

Fig. 7.—Field Connections.

armature. The series winding is used only with motors in automobiles.

A shunt winding is shown by the central figure in which the two ends of the field coil are connected directly across the two brushes, thus placing the armature and fields in parallel instead of in series. The two leads L-L' from the brushes B-B' lead directly to the external circuit. The field winding of the shunt type is of comparatively fine wire having a high resistance so that only a small amount of current will pass through the field. This winding is seldom used in either starting motors or lighting generators for automobiles but is often met with in stationary installations. Current from the positive brush B' leaves the brush at b, passes through e' and the fields F' and F, from which it returns through e and a to the negative brush B.

A compound wound motor has both shunt and series fields the connections of which may be easily traced from the lower figure in which FS and FS' are the shunt fields and F-F' are the series fields. The shunt fields are connected to the brushes B and B' at a and B', the two half fields being connected by the bridge e. One end of the series field F' is connected to the brush B' while the other end of the series field leads directly to the external circuit from L. The entire current passes through the series fields as before.

Nearly all starting motors are series wound since this type gives a better torque or drive than either the shunt or compound. This great torque is particularly noticeable in starting the motor from a standstill as the sudden rush of current passes through the fields as well as the armature, thus increasing the magnetic flux and the torque on the armature conductors.

The shunt winding tends to hold a fairly constant voltage in the generator through normal outputs, but with heavy drafts of current the voltage drops owing to the low resistance of the external circuit. Shunt wound motors tend to maintain a constant speed with a constant line voltage and varying load, but have a very low starting torque.

Compound wound machines may be divided into four principal groups, the long compound, the short compound, the cumulative compound and the differential compound depending upon the relations existing between the shunt and series fields. The "long" and "short" divisions are really subdivisions of the cumulative and differential groups since either may be made long or short. In general the dynamos used for automobiles are of the short differential order, a type that tends to maintain a constant current with a varying driving speed. In the cumulative type the shunt and series fields act together, an increase in either field causing an increase in the total magnetic flux. When the load increases the current naturally increases in the series field which in turn raises the voltage, hence with the cumulative compound an increase in current output causes a corresponding increase in voltage at the terminals of the machine.

In the differential compound the shunt and series fields oppose one another in such a way that an increase of current through the series winding tends to decrease the total magnetic flux. Since an increase in the current output thus diminishes the flux it follows that the voltage is also diminished, thus tending to maintain a constant flow of current. In practice it is possible to build a differential compound that will maintain a constant flow of current with widely varying speeds. This is the condition to be met in charging the cells of a starting and lighting battery. When the shunt fields of either the differential or cumulative compound are connected directly across the brushes the machine is known as "Short." When the shunt winding is connected to a brush at one end and to the outer end of the series field at the other the connection is known as a "Long" compound.

MOTOR-GENERATORS.

In some types of self-starting apparatus, notably the "Delco," a single unit performs both the functions of a motor and generator. This machine has a double wound-armature and is provided with two commutators, one commutator and winding for the generator function and one for the motor. A machine of this type is known to the automobile trade as a "motor-generator." This however is not the correct electrical term, since this sort of machine has been known as a "Dynamotor" for many years in electrical work.

Motor generators are always compound and are so connected to the switching and starting apparatus that they are alternately made cumulative and differential. To have a constant charging current the motor-generator is differentially wound when running as a generator, but is changed to a cumulative compound when running as a motor. This change is effected generally by reversing the current flow through the series field winding.

DELCO SELF-STARTING AND LIGHTING SYSTEM

With the exception of one model, the lighting, starting and ignition are all performed by one unit in the Delco system. The motor generator is provided with a double winding on the armature and has two commutators, one for the motor service and the other for the generator. A Delco timer and high tension distributer, together with the spark coil, are mounted on the motor generator. When running as a generator, the unit is driven through an overrunning clutch. When operating as a motor the armature drives the engine through a reduction gear which meshes with teeth cut in the engine fly-wheel rim through a clutch, so that the motor can be connected or disconnected from the engine at the will of the operator. It should be understood that it is the policy of the company to fit the system to the car rather than to market one single model to meet all conditions.

In Type "A" both commutators are mounted on the front end of the shaft. One motor brush and one generator brush are mounted on a common rocking support in such a way that when the generator brush is in contact, the motor brush is out of contact, and vice versa. A switch which regulates the flow of current from and to the machine is mounted on the brush support, so that both are operated by the starting pedal. When the pedal is depressed, the generator windings are disconnected from the circuit by the switch, the generator brushes are lifted, and the motor brushes placed in contact with the motor commutator.

When running as a motor the fields are cumulative compound wound, both a shunt winding and a reverse series winding being in circuit at this time, the series winding improving the torque of the

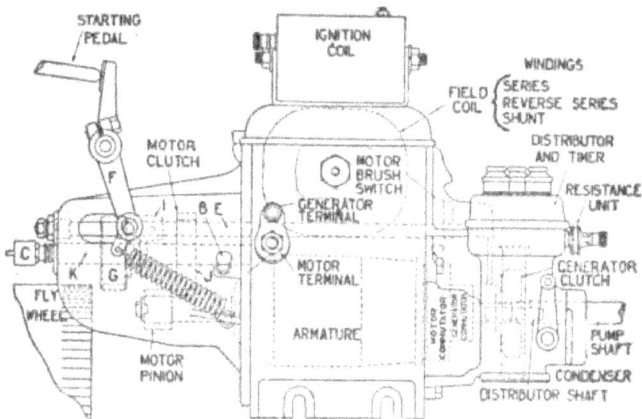

Fig. 8.—Delco Motor-Generator as Built for the Cadillac Lighting and Starting System.

motor. When running as a generator this winding with series reversed becomes a differential compound and maintains a constant voltage without the use of an external current regulator. All of the units are of the single wire type, the return current passing through the frame.

The main switch which controls the current between the generator and battery also controls the ignition system, so that no connection exists between the generator and battery until the ignition current is turned on at the switch. With the generator standing, this allows a little current to pass back through the generator between the time that the ignition is turned on. This reverse current from the battery turns the motor over very slowly, making it easier to mesh the gear with the teeth in the fly-wheel. The amount of current lost in this way is very small. This is called "motoring the generator."

CIRCUIT CONTROL SYSTEM

Two buttons are provided for the control of the ignition, one button (M) controlling the current from the battery to the circuit breaker and the other (B) controls the current from the auxiliary dry battery set. When either button is operated it not only closes the ignition circuit but also connects the generator with the storage battery, as noted above. An automatic circuit breaker is included in the main circuit for the protection of the apparatus in case of a short circuit or ground. This is a coil of wire wound on an iron core which, with an excessive current passing through the windings, makes and breaks the circuit intermittently in much the manner of a buzzer, thus giving warning of a fault in the circuit.

The ignition system is of the ordinary Delco type, except that there is a coil of resistance wire connected in the primary circuit between the primary of the spark coil and the timer. Under normal conditions this coil has little resistance and impedes the current to a negligible amount. Should the main switch be carelessly left open, however, with the motor standing still, the uninterrupted current will heat the coil, increase the resistance, and greatly reduce the flow of current through the breaker and spark coil. This will prevent damage to the spark coil and will save the storage battery from a rapid and injurious discharge.

CIRCUIT DIAGRAM OF DELCO SYSTEM

Referring to Fig. 9, the circuit diagram, the generator and motor are shown as two different instruments, though they are really combined. The upper contacts K and X on buttons M and B complete the ignition circuit and the lower contacts J and Z control the motoring of the generator. Upper and lower contacts operate together, on the same rod. The upper contact X on button B simply supplies current for the ignition from the dry battery.

Assume the conditions as shown, that is, all buttons down. Both dry and storage batteries are on open circuit. Current cannot go to the motor because the brush L is up. It cannot go to the generator, because the contacts J, K, X and Z are open. If the engine were being cranked by hand, with the ignition off as shown, the only passage of current would be from the generator armature through the brush A, reverse series field R to F to shunt field S to ground G-1, G-2 and brush H back to the armature.

Assume we get ready to start: we pull up button M, closing contacts at K and J. We get ignition current from the storage battery in the following way: From the plus terminal of the battery through E, switch terminal 1, D, circuit breaker Q, B, K, Y, switch terminal

7, primary of ignition coil P-1, resistance unit R-1 timer, to ground at
G-7—with an electrostatic ground through the condenser C-1 to G-5.
The circuit is completed through the ground to G-4 and the negative
terminal of the storage battery.

Motoring of the generator occurs when M is pulled out. In this the
circuit is to D, where some of the current goes to C, to the contact J,
to switch terminal 6, through the reverse series field ·R, to generator

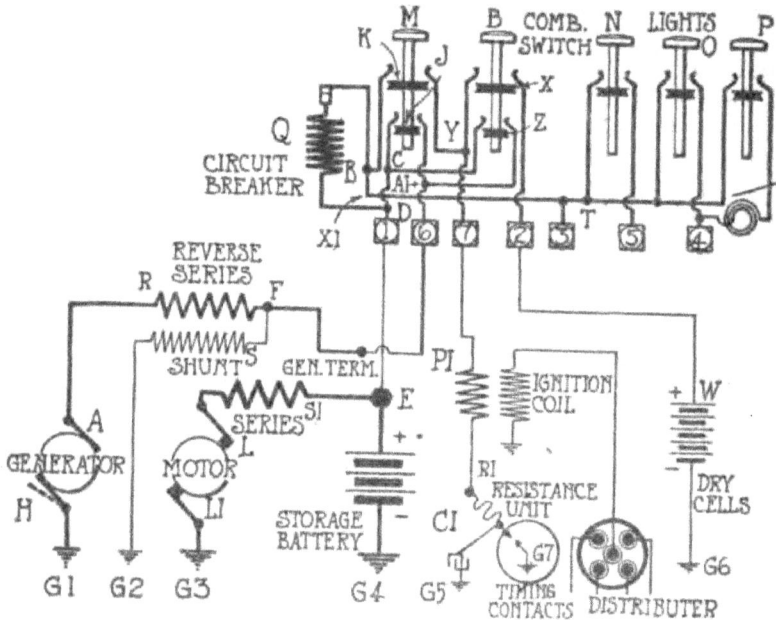

**Fig. 9.—Delco Lighting and Starting Circuit. (Courtesy of
"The Automobile.")**

brush A, to brush H and to ground G-1, completing the circuit
through ground G-4 and the battery. A part of the current divides
at F and goes through the generator shunt field to ground G-2. Current through the generator causes it to be driven slowly as a motor.

If button B were pulled up instead of button M, the connections
would be just the same, except that current from the dry battery
would be impressed on the ignition circuit from the positive terminal
of the dry battery, through 2, X to Y and the ignition circuit previously outlined. At the same time storage battery current would
be sent to the generator from the point C through Z to the generator
connection at A-1.

Now, with the ignition on and the generator running as a motor, we press the starting pedal. This shifts the pinion on the end of the armature shaft, which is rotating slowly, into mesh with the gear train to teeth on the fly-wheel. As the pedal completes its movement, linkages lift the generator brush H away from its commutator and lower the motor brush L to its commutator. This disconnects the generator armature from the circuit and throws the motor in. Storage battery current is then impressed directly on the motor through E, the series field S-1 brush L, the circuit being completed through brush L-1 to ground at G-3. This rotates the motor and through the fly-wheel starts the engine.

When the engine has taken up its cycle, the starting pedal is released, brush H goes to the commutator and brush L leaves its commutator, putting the generator back into circuit and cutting the motor out. Also the pinion is retracted from the fly-wheel, and the generator starts to furnish current for the battery and ignition. This is accomplished through the following circuit: A, R, F, 6, J, C, D, where it divides, part going to the battery through 1 and E to the positive terminal of the battery, and part through Q, S, K, Y, 7, to the ignition.

Current for the lights and horn comes from the battery or generator from S to the bus bar X1Y1. Button N supplies the sidelights V from T through U to 5. Button O supplies the headlights, and P throws in the dimmer resistance.

The Delco system, Type B, is the one employed on the greater number of cars equipped with this make of starting, lighting and ignition. There is a difference in the regulation of the generator output. It will be remembered that in the type already described the voltage regulation is obtained by the compound winding of the generator fields, having a shunt field and a reverse series field.

In the Type B, the reverse series field is omitted, leaving only the shunt field, and the regulation is obtained by an automatic regulating resistance, which also acts as an automatic spark control and is contained within the distributer housing. It consists of a spool of bare resistance wire, inserted in the shunt field of the generator. On this spool is a contact, which can move up or down on the spool. The contact is shorted direct to ground, so that as the contact moves up on the spool it cuts in more and more coils of resistance wire, and when it is at the top of the spool, all the resistance is in circuit and the current through the shunt field is weakened correspondingly.

The contact is attached to an arm operated by a centrifugal governor on the timer shaft, so that as the armature speeds up, more resistance is inserted and the output of the generator thus controlled. Also the ignition resistance is grounded through the regulating resistance and is cut out of the ignition circuit when the arm is at top position. This increases the intensity of the spark at high speeds.

In the C type the combination switch is of a different type and has not the double-contact feature of the others. Instead of this an automatic cutout relay is used to close the circuit between the generator and storage battery when the generator voltage is high enough to charge.

Instead of the governor-controlled variable resistance used in the Type B systems to control the generator output, a solenoid mercury well voltage controller is employed. Motoring of the generator is obtained by a push button on the combination switch.

There is an added feature in the ignition circuit of the Type C outfits. This is the ignition relay which is connected in the dry battery circuit. It serves to break the primary circuit immediately after it is completed by the auxiliary contact mechanism in the timer, thereby inducing a high tension flow through the secondary circuit, which results in a hot spark in the spark plug. It is simply an electromagnet, having two windings, one of coarse wire carrying the main ignition current. It is connected so that current through this winding energizes the core, attracts an armature which opens the circuit. The other winding is of comparatively fine wire connected around the point of break in the main circuit, so as to hold the armature down by the current shunted through the fine winding when the main circuit is opened. If it were not for the fine wire winding the making and breaking of the main circuit would give a vibrating spark at the plug, instead of a single spark. This vibrating effect is purposely produced for starting by opening the circuit through the fine wire winding, when the button on the combination switch is held in.

BIJUR STARTING SYSTEM

Either a motor-generator or an independent motor and generator are used in the Bijur system according to the conditions. The ignition apparatus is not included in the Bijur circuit, this being installed separately by the automobile builder according to his judgment. The motor generator drives or is driven through a silent chain to the crank case. The two unit system of a motor and generator start the car and supply current for the starting and lighting, the generator being of the constant voltage type. The voltage is held constant at any engine speed by means of an automatic regulator that varies the current flowing in the field magnets.

With a constant voltage it is possible to obtain a tapering charge that is to have the current flow heavier at the beginning of the charge than at the end, a very desirable feature in the proper maintenance of a storage battery. Since the voltage of a partially discharged battery is lower than the voltage at full charge, there is less opposition to the flow of generator current and a higher charging rate is the result. As the charge continues the voltage of the battery rises and cuts down the 'current, this "tapering" down the charge. The system can be wired either for the one wire (grounded return) or the two wire circuit.

In general the circuit for light can be considered as a battery lighting circuit with the generator placed in parallel with the battery, and all of the apparatus, the lights, horn, ignition, and starting motor circuit all receive current from the battery terminals. Any of the wires can be connected with any piece of apparatus without regard to polarity, not even the generator being affected by the polarity of the wires leading to it from the battery.

Gear drive to the fly-wheel is used with the independent motors, the pinion being thrown in and out of mesh with the fly-wheel in a manner similar to that of the Delco, an intermediate gear train being used for reducing the motor speed. In another type of independent motor drive the motor acts directly on the fly-wheel through a pedal operated pinion. This motor is built to stand high speed and will not be injured if left in mesh after the engine starts normal firing.

BIJUR CIRCUIT DIAGRAM

While the circuits may differ slightly in detail, Fig. 10 will give the general principles of the Bijur circuit, in which (B) is the battery, (G) is the motor-generator, (A) is the ammeter, (C) is the temperature regulator, (R) is the cut-out relay, (L) is the magnetic latch, (V) is the voltage regulator, (D) is the main knife switch, and (E) is the starting button.

It will be noted that the motor-generator (G) is provided with an armature having a double commutator (1) and (2), a single pair of field coils (F) and (F¹), and a double wound armature (H).

The field magnets of the motor-generator are compound wound, that is to say, that the winding shown by the thin lines is in parallel with the brushes while the coil, shown by the heavy lines, is in series with the main current so that all of the current passes through the series field. Since the voltage of a generator increases with its speed it is necessary (in order to maintain a constant voltage) to decrease the current through the shunt field by means of an adjustable resistance such as the voltage regulator (V). This regulator is connected to the shunt field by the wire (4) and to the opposite side of the circuit by the wire (14).

This adjustment of the field resistance is accomplished magnetically by the magnet coil (I), which acts on an iron plunger within the case. The plunger carries a coil of resistance wire in such a way that it may be drawn in and out of a mercury bath in the bottom of the regulator, thus varying the length of wire in circuit. Should the voltage rise slightly above normal, the magnetic pull of the coil (L) will be increased, thus raising the plunger farther out of the mercury and increasing the resistance of the field circuit. If the engine speed should fall the plunger will drop, cutting out the field resistance in proportion to the fluctuation in the voltage, thus increasing the field current and raising the generator voltage again to normal.

A cut-out relay (R) is provided with a double winding (8) and (9),

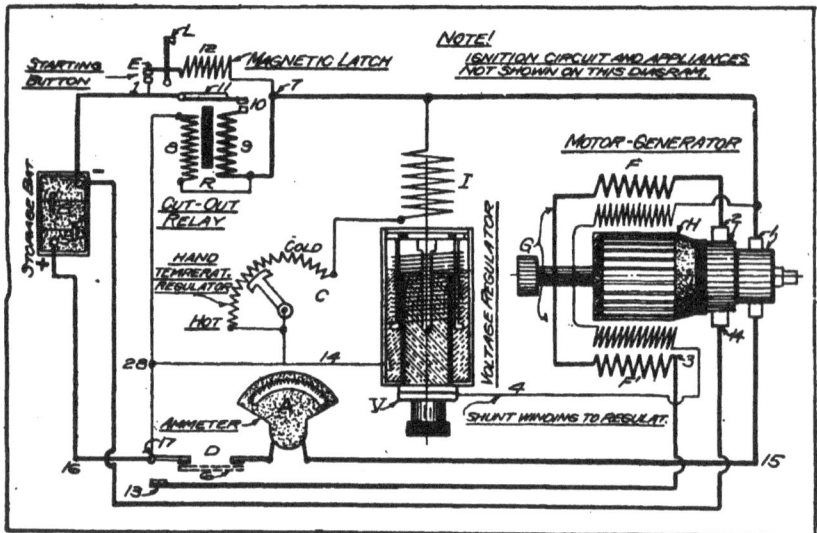

Fig. 10.—Circuit Diagram of Bijur System.

the first mentioned being a shunt winding while the latter is in series with the charging current. These windings are on an iron core which act on the spring controlled armature (11). This armature actuates the contact points (10) which open and close the charging circuit. When the voltage of the generator is above that of the battery, both windings act together in holding the contacts together and closing the circuit. Should the generator voltage become less than the battery, the flow passing through the series winding (9) will be reversed, a condition that will neutralize or kill the magnetic field in the relay core, allowing the spring to open the contacts and break the circuit.

A magnetic latch (L) with the magnetic coil (12) is connected in the starting circuit in such a way that on completing the circuit with push button (1) the coil will draw the latch back, connecting the starting pedal and gear train together mechanically. It is impossible to operate the gears before this is done.

Because of the variations in the resistance of the storage battery due to temperature changes, an adjustable ballast resistance (C) is in series with the regulator magnet (I). This is moved manually and controls the voltage limits.

OPERATION OF BIJUR STARTER

On pushing the starting button (1), the current flows from the battery terminal (5) across the switch (6), through the ammeter (A) and into the motor commutator (1). This runs the motor at about 100 revolutions per minute for meshing the starting gears. From the other brush on the commutator the current flows through the coil (12), through the contacts (1) and back to the negative battery terminal (7). No current can flow through the cut-out as the contacts (10) are open when starting. A part of this current flows through the voltage regulator windings (I). In passing through the coil (12) of the magnetic latch the armature is drawn back, connecting the starting pedal and the gear shift.

The starter pedal, on being pushed out, first meshes with the gears on the armature shaft and then, with the teeth on the fly-wheel, the slowly revolving armature making this last a simple matter. When the pedal is full out a switch is actuated by the pedal that gives the full current to the motor. The switch (6), on being opened at this moment, disconnects the original starting button current, the current now flowing from the battery terminals through the contact (13), through the series field (3) of the motor, through the armature at (2) and back to the storage battery negative pole (7). This cranks the motor at full speed and when the motor begins to fire the pedal should be released quickly, the release throwing the gears out of mesh and releases switch (6).

With the engine firing, the generation of current begins. The current flows from the commutator at the right, through the ammeter to (16) and thence to the positive pole of the storage battery. No amount of current can flow through the battery since the contacts (10) are open in the main circuit, but a small amount does flow through the relay shunt coil (8), causing the armature to be drawn down and the main circuit closed when the voltage of the generator reaches a value of about 7 volts. With the main circuit closed the charging current from the battery now passes through the series coil (9), which aids in keeping the points in contact.

Should the current become reversed through (9) owing to a drop of voltage or speed in the generator, the polarity of (9) will be reversed and will act against the field of the magnet (8), thus killing the magnetism and allowing the spring to pull up the armature and open the circuit.

GRAY-DAVIS SELF-STARTING SYSTEM

The greater part of the self-starting and lighting equipment made by the Gray-Davis Company is of the separate unit system, the motor, generator and ignition apparatus being entirely independent mechanically. In one type only the distributer and generator are in one piece.

In the older type of generator the voltage was held constant by the use of a centrifugal governor, but in the later models the voltage is

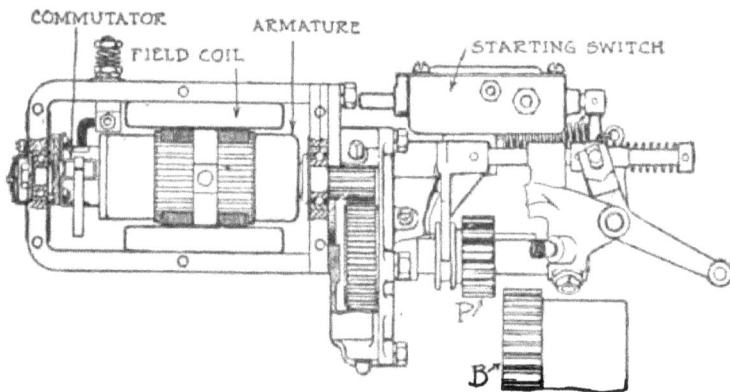

Fig. 11.—Gray-Davis Motor Equipment in Which P is the Motor Pinion and B the Fly-Wheel Gear. The Shifting and Starting Lever is at the Right. (Courtesy of "Motor Age.")

maintained electrically by a device placed on the top of the generator. A cut-out disconnects the generator when its voltage falls below that of the battery. The generator designed for four cyclinder cars runs approximately at $2\frac{1}{2}$ times the crank-shaft speed. The six cylinder model runs at cam-shaft speed, or $1\frac{1}{2}$ times crank-shaft speed. This allows the generator to run at its rated speed of 650 revolutions per minute, with a car speed of from 10 to 12 miles per hour.

Fly-wheel drive is standard, with the reducing gears, starting switch and gear shift mounted as an integral part of the starting motor. The movement of the pedal closes the starting switch and meshes the gear with the fly-wheel. In a model brought out for use in the Ford car the motor and generator are arranged in the "double deck" method; that is, one unit is placed above the other, both, however, being contained in a single casing.

The Model T. generators are rated to give an output of 10 amperes at 6.5 volts, with a speed of 1,000 revolutions per minute, while Model S gives 10 amperes, 6.5 volts, at 650 revolutions per minute. The Model T generator weighs 20½ pounds.

Type Y motor develops full torque with 100 amperes at 2,800 revolutions per minute. For very heavy cars Type H-1 motor is used, this pulling full load at 1,500 R. P. M., with a current of 150 amperes at 6 volts.

THE ECLIPSE-BENDIX STARTER DRIVE

This device is an attachment for the starting motor which does away with the necessity of an overrunning clutch or a starting pedal, and can be applied to any starting motor that drives through the fly-wheel.

Generally speaking, it consists of a threaded shaft on which a threaded pinion is mounted, the shaft being connected with the armature shaft of the motor by means of a coiled spring. A weight is attached to one side of the pinion so that it is out of balance, the weight normally hanging down. This pinion may now be considered as out of mesh with the teeth of the fly-wheel.

When the starting motor starts to revolve, the pinion is pulled along the shaft toward the fly-wheel teeth, since the pinion is prevented from turning by the counterweight. This action is similar to the travel obtained by holding a nut on a revolving screw, since the hole in the pinion is threaded like a nut. It is pulled across the entire face of the gear until it meets a collar on end of the shaft, at which time, of course, it starts revolving with the shaft and drives the fly-wheel. The impact of this sudden termination of its travel is taken up by means of the spring that connects with the armature shaft so that no shock is transmitted to the motor.

When the motor starts firing and comes up to speed it runs away from the pinion, turning it in the opposite direction so that it runs along the shaft and out of mesh with the fly-wheel. In other words the pinion is "unscrewed" on the shaft, placing it in its former position.

As the weight is now revolving with the pinion at a high speed, the centrifugal force exerted by the weight causes it to bend on the threads of the shaft so that continued rotation of the motor will not cause it to travel back into mesh.

BOSCH-RUSHMORE STARTER SYSTEM

The Bosch Magneto Company market two self starters, the Bosch, which is a system of their own, and the Bosch-Rushmore, which was made by the Rushmore Dynamo Company.

The Bosch-Rushmore consists of a separate generator and motor with an independent magneto, the motor driving the fly-wheel through an arrangement that automatically engages and disengages the armature shaft pinion with the teeth on the fly-wheel.

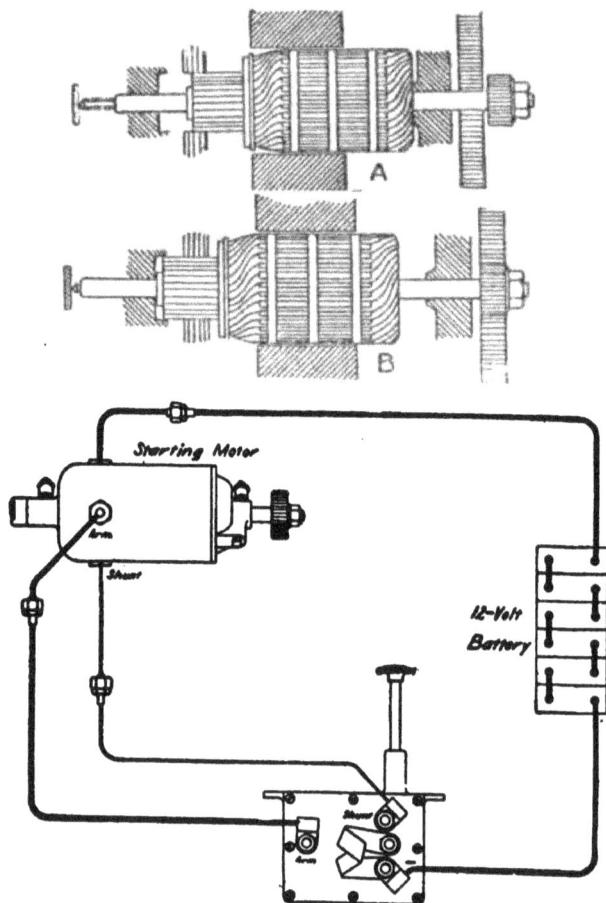

Fig. 12.—Above is Shown the Shifting Armature Device of the Bosch-Rushmore System. At A the Armature is Out of the Field Giving a Low Voltage. At B the Generator is Giving Full Voltage With Armature Central in the Field. The Spring is in Shaft at Left of Armature.

A spiral spring placed on the commutator end of the motor shaft which forces the armature to one side of the pole pieces, the armature shaft being constructed so that it can slide longitudinally in its bearings. When current is passed through the motor, the armature is drawn magnetically under the poles against the compression of the spring, and at the same time the motor pinion is drawn with the shaft and into mesh with the fly-wheel gears. As the armature is now under the poles it can exert its maximum torque.

In the Bosch self-starting system there is a separate motor and generator, the motor usually being connected with the crank-shaft through a silent chain, an over-running clutch allowing for the speed difference due to the two extreme running conditions. One generator is of the magneto type having a permanent magnetic field and is designed to run at engine speed. The other Bosch generator is a shunt wound machine having an automatic field current regulator located in the dash switch-board.

Starting is effected by a relay circuit operated through a push button, the closing of the relay circuit closes the main switch magnetically and starts the motor. When the starting handle is retained, the insertion of the handle connects the battery to the ignition system and cuts out the magneto. When the engine starts firing, the battery is automatically cut out and the magneto is replaced in the circuit.

AUTO-LIGHT CRANKING SYSTEM

The two unit auto-light system operates on 6 volts both for the cranking and lighting systems. A conventional circuit breaker, pedal switch, and storage battery are placed in the usual circuit, the breaker cutting out the generator to prevent a reversal of the current. In one model a primary circuit breaker and distributer for the ignition system is mounted on the generator. Drive can either be to the crank shaft by silent chain, to the fly-wheel through a gear train or to the transmission. Due to the construction, the units can be mounted either horizontally or vertically. The usual gear reduction from the motor is 25 to 1.

A differential compound winding maintains a constant current. Charging begins at an engine speed of about 200 revolutions per minute, which corresponds to a car speed of about 5 miles per hour. The output increases with the car speed until it reaches 17 miles per hour, at which point the current is 12 amperes. At this point the series winding holds the output constant so that it is no longer affected by an increase in speed. The generator operates at engine speed so that it is possible to drive from the magneto shaft.

The model M cranking motor is of the series wound type weighing 36½ pounds. It is claimed that this motor will crank a six cylinder engine under 60 pounds compression at 100 revolutions per minute, the current draft being 95 amperes.

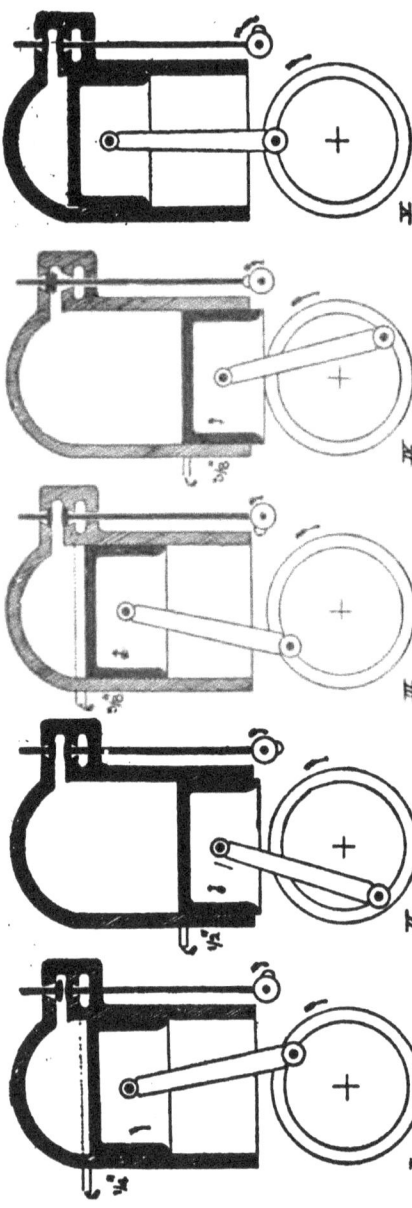

I

SUCTION STROKE.
Inlet valve *opens* ⅛in.
after piston has started
to descend.

II

COMPRESSION
STROKE.
Inlet valve *closes* ⅝in.
after piston has com-
menced to ascend.

III

COMPRESSION
STROKE
Spark contact is made
⅜in. *before* piston has
reached dead center,
with ignition fully ad-
vanced, after which
downward power stroke
takes place.

IV

EXHAUST STROKE.
Exhaust valve *opens* ¾
in. *before* piston com-
mences to ascend

V

EXHAUST STROKE.
Exhaust valve *closes*
when piston has
reached the dead
center.

The dotted lines in all cases denote the dead center at each end of the stroke.
The times of opening and closing the inlet valves only refer to those valves which are mechanically operated.
In this diagram the positions of the piston are shown at the time when the inlet valve opens, when it closes, when the
sparking should take place, the exhaust valve opens, and when it closes.   Different makers adopt different
timings, but above is about the average of 14 well-known makers.

The timing of the spark contact refers only to high tension ignition. When the contact is broken inside the cylinder,
as in the low tension magneto system, it should be broken almost immediately on the dead center.
N.B.—This diagram shows automatically operated inlet valves.   The timing will be the same with mechanically operated
inlet valves.

The Four Stroke Cycle.

PART X

VALVE SETTING AND TIMING

To understand the "timing" of motor valves one must have the principle of the four cycle motor in mind. The periods of valve opening and closing vary widely in different types and makes of motors. With equal areas of pipe and valves, and with a constant suction, less mixture will enter the cylinder of a high speed motor than in a low speed type, since the rate of gas flow will be equal in both cases. In other words, the gas will not have time to completely fill the cylinder of the high speed motor before the piston reaches the end of the stroke with identical valve timing.

In general there are two methods of increasing the rate of gas flow: (1) By increasing the area of the valve opening; (2) by opening the valve earlier and closing it later, thus increasing the time period of flow.

Both methods have limitations, since for certain practical reasons it is impossible to increase the valve area beyond certain sizes, and as we have certain functions to perform in predetermined angles we cannot increase the time of opening beyond a certain point. Increasing the valve diameter beyond a well defined limit will result in valve warping, or in excessive wear on the cams, push rods and valve seats due to the increase in weight. Owing to the increased inertia of a heavy valve we will require heavier valve springs which still further increases the wear on the moving parts.

Lifting a valve higher to increase the area of opening still further increases the stress and wear on the valve actuating mechanism since the stress varies as the square of the velocity. A high lift causes the valve to hammer on the seat with unnecessary violence and is a prolific cause of noise and vibration. To actuate a heavy valve against a heavy spring pressure at high speed absorbs no inconsiderable percentage of the motor's output of power. The exhaust valves are the worst offenders in respect to wear and power loss since they open before the end of the stroke or before the gas has expanded to a low pressure. The earlier the exhaust valve is opened the higher will be the gas pressure and the greater the wear. A very early exhaust opening also decreases the output of the motor since the gas has not exerted its pressure as far in the stroke as it should. Delaying the inlet valve closure beyond a certain point materially shortens the compression stroke and interferes with the combustion chamber and valve pockets.

Long manifolds or intricate passages with short sharp bends increase the gas friction and therefore limit the rate at which the gas flows. The time given for the gas under a slight suction pressure is extremely short even in low speed engines.

Since all motors vary in regard to valve areas, speeds, lifts, length of passage, etc., it will be seen that the only true authority for any particular type of motor is the manufacturer. For this reason we give a list of timings advocated by a number of prominent automobile manufacturers. The use of this table in connection with the directions for making the adjustments will enable the reader to set the valves of almost any modern American automobile. In the case of cars using a stock motor, determine the make and model of motor and consult the maker's name in the list.

THE FOUR STROKE CYCLE (First Revolution)

The four stroke cycle motor, or incorrectly called the "four cycle motor," accomplishes all of its operations in two revolutions or four strokes. There is one power impulse in every two revolutions. On the first stroke (suction stroke) the gas is drawn into the cylinder through the inlet valve by the piston. During this stroke the inlet valve remains open. The exhaust valve must remain closed so that a sufficient vacuum can be produced for the movement of the mixture. At, or a little past the end of the stroke, the valve closes.

COMPRESSION STROKE (First Revolution)

The piston now returns to the outer end of the stroke with both valves closed, compressing the gas before it.

POWER OF EXPLOSION STROKE (Second Revolution)

Both valves remain closed until a little before the piston reaches the end of the stroke, at which point the exhaust valve opens and allows the spent gas to escape to atmosphere.

SCAVENGING STROKE (Second Revolution)

The piston moves outwardly expelling the burnt gas through the exhaust valve which remains open through the entire stroke. The inlet valve remains closed.

At the completion of the "scavening stroke" the entire cycle is repeated. The exhaust valve usually remains open slightly after the end of the scavenging stroke. In the following description the term "moving outwardly" means that the piston is moving away from the crank-shaft. As practically all modern automobile motors are of the vertical type the term "outwardly" would have the same meaning as "upwardly."

ANGULAR MEASURE OF EVENTS

The time at which any event takes place, such as the opening or closing of a valve, or its duration, is usually measured in angular

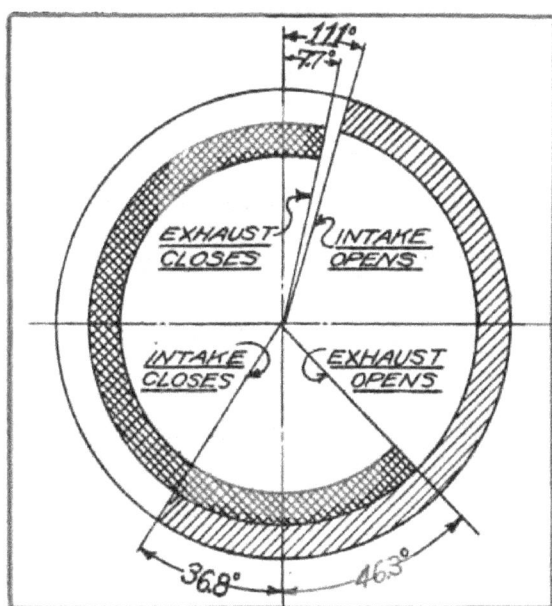

Fig. 1.—Valve Setting Diagram.

degrees taken on the crank circle. For example.—If the inlet valve is open for 200°, we mean that the crank will swing through an angle of 200° from the time that the valve starts to open until it is closed. In locating the point at which a certain event occurs we refer to either the upper or lower dead center. For example.—When the inlet valve of a certain motor opens 11° past the upper dead center, we mean that the crank will move from the upper dead center through an angle of 11° before the valve opens. All valve settings are usually shown by diagram such as Fig. 1. In this particular figure the outer cross hatched ring represents the length of time that the inlet valve is open. The inner ring represents the angle through which the exhaust valve remains open. The intersection of the vertical center with the outer circle at the top of the diagram represents the upper dead center. The upper inclined dotted line is the position of the crank center at the time the inlet valve opens, which is shown as 11.1°. Following the outer ring around to the bottom of the diagram we find by the inclined dotted line that the inlet closes 36.8° after the lower dead center. In figuring the total opening of the inlet valve we find it to be (180 — 11.1) + 36.8 = 205.7°. In this case we see that the inlet valve opens after the upper dead center and closes after the lower dead center being open more than the 180° that we have considered in describing the theoretical cycle of the engine.

From the lower inclined solid line we see that the exhaust valve opens 46.3° before the crank reaches the lower dead center, and closes 7.7° after the upper dead center. The length of exhaust valve opening is therefore 7.7 + 180 + 46.3 = 234°. In the theoretical cycle this would have been 180°. This diagram represents the average of a great number of 1915 four cylinder models, the data being compiled by the staff of "The Automobile." According to the same authority the average timing for 1914 four cylinder models was as follows:

> Intake opens 11.2° past upper dead center.
> Intake closes 35.0° past lower dead center.
> Exhaust opens 50.0° before lower dead center.
> Exhaust closes 9.3° past upper dead center.

The timing of the average six is slightly different than that given for the four (1915 models):

> Intake opens 10.7° past upper dead center.
> Intake closes 37.6° past lower dead center.
> Exhaust opens 46.0° before lower dead center.
> Exhaust closes 7.0° past upper dead center.

In comparison with the fours and sixes given above we will give the timing of the 1915 eight cylinder Cadillac motor (high speed type). Bore and stroke = 3⅛ x 5⅛ :

> Intake opens 0.0° past upper dead center.
> Intake closes 46.6° past lower dead center.
> Exhaust opens 46.6° before lower dead center.
> Exhaust closes 0.0° after upper dead center.

It will be noted on this high speed motor that the inlet and exhaust open and close respectively on the upper dead center, and that the inlet valve remains open way past the lower dead center in order to give the gas time to enter. The total period of inlet opening is 226.6°.

In reading the diagrams it should be noted that rotation of the crank is assumed to be toward the right or in the direction of the hands of a clock. Hence, if an event is said to be after upper dead center, it will occur after the crank has passed the center when turning in a right-handed direction.

VALVE LAP

We have seen that the exhaust valve closes and the inlet opens after dead center, but that these events do not usually take place at the same time, the exhaust closing first. The exception to this rule is the Cadillac 8, which in following aeronautic practice opens the inlet and closes the exhaust simultaneously—on the upper dead center.

The angle between the closing of the exhaust and the opening of the inlet is called the "valve lap," and is expressed in degrees. In the average 1915 four cylinder motor this would be from diagram, $11.1 - 7.7 = 3.4$ lap. The Cadillac is an example of zero lap since the opening of the inlet and the closing of the exhaust takes place at the same time.

There are three conditions of lap, negative, positive and zero. With the positive lap both valves are open together, in negative lap the exhaust closes before the inlet opens, and with zero both actions take place at the same time. With negative lap the piston descends before the inlet opens creating a slight vacuum, increasing the rush of gas into the combustion chamber. With zero lap there is no vacuum at the entrance point. With positive lap the inertia of the exhaust gases creates a slight vacuum which is an aid in forcing the mixture into the cylinder. This is generally used with aeronautic and tee head type automobile motors, as the incoming and outgoing gases do not conflict.

AUXILIARY EXHAUST PORTS

To decrease the amount of hot gas passing over the exhaust valve, auxiliary exhaust ports are sometimes provided. These are similar to the exhaust ports cast into two-stroke cycle motors. The ports consist of either a series of holes cored or cast in the cylinder walls and located so that the piston top uncovers them at the inner end of the working stroke.

TIMING TABLE—LATEST MODELS

Car Name	Model	Inlet Opens Deg. Past Top Center	Inlet Closes Deg. Past Bottom Center	Exh. Opens Deg. Before Bottom Center	Exh. Closes Deg. Past Top Center
Abbott	F-P	10	28	40	2.5
	L	17	29	42	8
	K	11	44	45	11
Auburn	36	21	30	44	10
	6-40	21	39	43	12
	6-47	17	47	50	13
	43	5–30	36	71	15
† Beaver	K	10	35	55	5
	ML	10	30	45	5
	6A	10	25	38	8
	6B	10	25	38	8
	4A	10	25	38	8
	4B	10	25	38	8
	E	10	35	42	8
	N	10	30	45	5
Briscoe	A	0	45	47	5
† Buda	M	15	33	53	12½
	C	5	45	55	5
	T	5	45	55	5
	Q	5	45	55	5
	OM-3	5	45	55	5
	TM-3	5	45	55	5
	QM-3	5	45	55	5
	OU	5	45	55	5
	TU	5	45	55	5
	QU	5	45	55	5
	R	5	45	55	5
	RU	5	45	55	5
	V	5	45	55	5
	VU	5	45	55	5
	SS	5	45	55	5
	SSU-3	5	45	55	5
	SSU-4	5	45	55	5
	LS	5	45	55	5
	LSU	5	45	55	5
Buick	C-24, C-25	16–11'	35–41'	56–19'	13–11'
	C-36, C-37, C-54, C-55	14–5'	36–25'	56–51'	11–29'
Cadillac	8	0	46–40'	46–40'	0
Cartercar	9	15	38	45	10·

TIMING TABLE—LATEST MODELS

Car Name	Model	Inlet Opens Deg. Past Top Center	Inlet Closes Deg. Past Bottom Center	Exh. Opens Deg. Before Bottom Center	Exh. Closes Deg. Past Top Center
Case	40	13	30	50	13
	35	0	30	45	0
	25	10	30	40	5
Chalmers	26B	12	33	55	12
	29	12	33	55	12
Cole	Four	15	38	45	10
	Light Six	15	38	45	10
	Big Six	15	38	45	10
† Continental	6F, 6A, 6-C, 6-N, 6-P	10	28	40	2–30'
	C-R	11–30	44–12'	45–48'	11–30
	E.J.N.T.	17–53'	29–25'	42–36'	8–20'
† Davis	SB	10	35	55	5
	MB	10	27	47	5
	FA	10	30	47	5
	Y	10	30	47	5
Dorris	1-A-4	10	30	45	15
Ford	T	12	50	37	Closes on top
Franklin	6-30	8	33	51½	17
Glide	30	15	38	45	10
Haynes	30	5	35	47	2
	31	5	35	48	3
	32	5	35	48	3
† Hazard	C	11	35	45	3
	CX	11	35	45	3
	D	14	30	44	8
Herff-Brooks	4-40	On top cent.	34	50	5
	6-50	"	34	50	5
† Herschell	4404-A	14½	40	44½	10
	4402-J	2	40	45	On center
	4403-N	8	Before 45	45	8
	4401-J	2	40	45	Center
	4301-E	2	40	45	Center
	4001-M	8	40	45	Center

TIMING TABLE—LATEST MODELS

Car Name	Model	Inlet Opens Deg. Past Top Center	Inlet Closes Deg. Past Bottom Center.	Exh. Opens Deg. Before Bottom Center	Exh. Closes Deg. Past Top Center
† Herschell	4101-S	8	Before 45	45	8
	4405-C	2	40	45	Center
	4201-B	2	40	45	Center
	6201	2	40	45	Center
	6301	2	40	45	Center
Hupmobile	K	11	43	38	6
	H	25	35	40	20
Jackson	Olym. 46	15	38	45	10
	48-Six	15	38	45	10
Jeffery	93-2	18	46	47	15
	104	15	50	45	10
	106	18	46	47	15
King	C	9–43'–40"	30–38'–20"	32–10'–20"	5
Kline	5-pass.	5	45	55	5
	7-pass.	5	45	55	5
Krit	O	8	31½	39	2
	M	8	31½	39	2
Lex.-Howard	6M	10	28	40	2½
	6L	10	28	40	2½
	4K	10	20	44	10
Locomobile	L-4	Top cent.	46–22	50–52'	16–27' Aft. bot. cent.
	M-5	Top cent.	48–30'	56–10'	15–39'
Lyons-Atlas	K-4	5	36	63	5
McFarlan	T	10	32	47	10
	X	10	32	47	10
Marmon	41	19	40	56	12
	48	16	40	56	12
Maxwell	25	6	32	43	6
Metz	22	7	40	44	7
Moline		20	50	50	5

TIMING TABLE—LATEST MODELS

Car Name	Model	Inlet Opens Deg. Past Top Center	Inlet Closes Deg. Past Bottom Center	Exh. Opens Deg. Before Bottom Center	Exh. Closes Deg. Past Top Center
National	AA	5	45	55	5
† Northway	39	15	38	45	10
	40	15	38	45	10
	47	15	38	45	10
	49	15	38	45	10
	30	15	38	45	10
Oakland	37	15	38	45	10
	49	15	38	45	10
Oldsmobile	42	15	38	45	10
	55	15	38	45	10
Paige-Detroit	6-46	10	28	40	2½
	4-36	11–20'	40–26'	51–18'	11–40'
Partin	20	18–14'	52–28'	56–43'	12–55'
Peerless	48-6	8–40'	30–21'	43–53'	3–55'
	54	17–53'	29–25'	42–36'	8–20'
	55	17–53'	29–25'	42–36'	8–20'
Premier	A	10	40	40	10
Pullman	6-48	10	28	40	2½
R. C. H.	K	18–12'	52	35	11–30'
Regal	D	10	38¾	46¼	5
	D-1915	10	38¾	46¼	5
Reo	R & S	17–46'	36–25'	53–18'	14–13
† Rutenber	38	15	50	50	10
	28	18	46	47	15
	27	18	46	47	15
	22	15	50	45	10
Saxon	B	12	45	55	12
Scripps	8	18–53'	47–31'	43–25'	15–45'
Simplex	30 H. P.	10–59'	33–12'	56–52'	8–13'
	50 H. P.	13–40'	36–4'	60	16–26'
	75 H. P.	On dead center	41–51'	66–24'	16–26'
Speedwell		10 Before up dead center	30	46	1.6

TIMING TABLE—LATEST MODELS

Car Name	Model	Inlet Opens Deg. Past Top Center	Inlet Closes Deg. Past Bottom Center	Exh. Opens Deg. Before Bottom Center	Exh. Closes Deg. Past Top Center
Stearns	SK-4	4	40	60	At center
	SK-6	4	40	60	At center
	SK-L4	8	40	60	4
Studebaker	Six	12–30'	32–30'	45	7–30'
	Four	12–30'	32–30'	45	7–30'
	EC	12–30'	32–30'	45	7–30'
	SD	12–30'	32–30'	45	7–30'
Velie	15	10	28	40	2½
	14	10	28	40	2½
	12	7 Ahead	36	43	12
Willys-Overland	80	8	38	46	15
	81	8	38	46	15
	82	10	28	40	2½
Winton	21	20–45'	35–30'	54–40'	15–30'
† Wisconsin	A	15	45	45	10
	B	15	45	45	10
	C	15	55	45	10
	D	15	45	45	10
	E	15	55	45	10
	F	15	30	45	10
	H	15	35	55	5
	K	15	30	45	10
	M	15	45	45	10
	P	15	30	45	10

* Given in inches on the fly-wheel.
† Stock motors built for assembled automobiles.

TIMING TABLE—EARLIER MODELS

Car Name	Inlet Opens Past Top Center	Inlet Closes Past Bottom Center	Exh. Opens Before Bottom Center	Exh. Closes Past Top Center
Abbott 34-40	11–30	44–12	45–48	11–30
Abbott 44-50	17–53	29–25	42–36	8–20
Abbott Belle Isle......	10–00	28–00	40–00	2–30
Allen 40	15–00	40–00	45–00	10–00
Cadillac 1914 .... 4–20 to	14–20	36–26	31–34 7–00 to	17–00
Pathfinder 4............	11–30	44–12	45–48	11–30
Pathfinder Big 6........	10–00	28–00	40–00	2–30
Pathfinder Little 6......	10–00	28–00	40–00	2–30
Chalmers	12–00	33–00	55–00	12–00
Chandler Six	14–00	39–00	49–30	12–00
Chevrolet C	13–00	49–00	47–00	9–00
Chevrolet H2-H4	16–48–½	54–8–½	27–13	14–6–½
Case 40	13–00	30–00	50–00	13–00
Franklin M No. 4......	8–00	33–00	51–30	17–00
Haynes 26-27	5–00	35–00	47–00	2–00
Haynes 28............	5–00	35–00	37–00	2–00
Hupmobile 32.........	21–00	28–00	46–00	16–00
Oldsmobile 54	15–00	38–00	45–00	10–00
Jeffery 6-96	18–00	46–00	47–00	15–00
Jeffery 4-93	18–00	46–00	47–00	15–00
King B	9–44	30–38	32–10	5–00
Krit M/K-M-KR	12–00	28–00	39–00	2–00
Lyons-Knight K	10–00	40–00	60–00	on
McFarlan Six T........	10–00	36–00	43–30	10–00
Maxwell 4-35	5–00	40–00	35–00	on
Maxwell 4-25	6–00	32–00	43–00	6–00
Moon Six 50..........	10–00	28–00	40–00	2–30
Moon Four 42.........	14–00	24–00	31–00	21–00
Marathon, Winner, Runner, Champion	12–00	45–00	46–00	7–00
Jackson, Olympic 40, Majestic, Sultanic	15–00	38–00	45–00	10–00
S & M	10–00	28–00	40–00	2–30
Pratt 50	12–00	45–00	45–00	10–00
Paterson 32 and 33.....	15–00	38–00	45–00	10–00
Palmer Singer Brighton Six L	6–00	40–00	45–00	5–00
Paige Detroit 36.......	9–40	32–30	41–50	11–40
Paige Detroit 25.......	9–40	32–25	40–30	12–00
Republic E	15–00	30–00	45–00	10–00
Reo the Twelfth	18–00	36–00	53–30	14–00
Simplex 38	10–20	31–40	54–20	7–50
Simplex 50	13–10	34–40	57–30	15–40
Selden 49	13–00	26–30	48–30	7–30
Velie, 9-45, 6-40, 9-5, 9-4, 9-2, 6-5, 6-4.........	7–00	36–00	43–00	12–00
Velie 11-35	5–00	31–00	39–00	13–00
Speedwell A B C.......	10–00	28–00	40–00	2–30

	Cylinder No. 1	Cylinder No. 2	Cylinder No. 3	Cylinder No. 4
Firing Order: 1-2-4-3	Firing.	Compression.	Exhausting.	Intaking.
Firing Order: 1-3-4-2	Firing.	Exhausting.	Compression.	Intaking.

Crank Arrangement for 4 Cylinders.

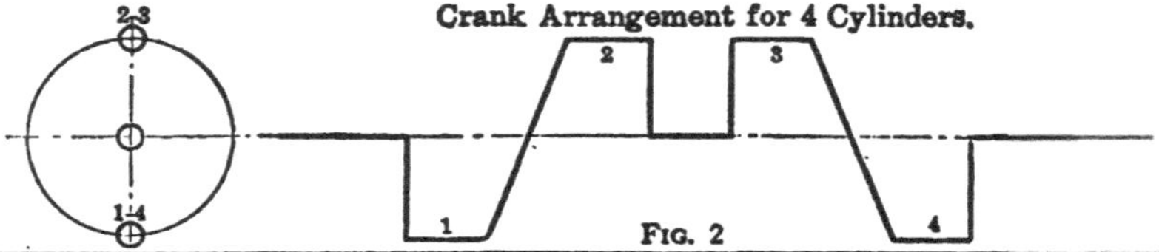

Fig. 2.

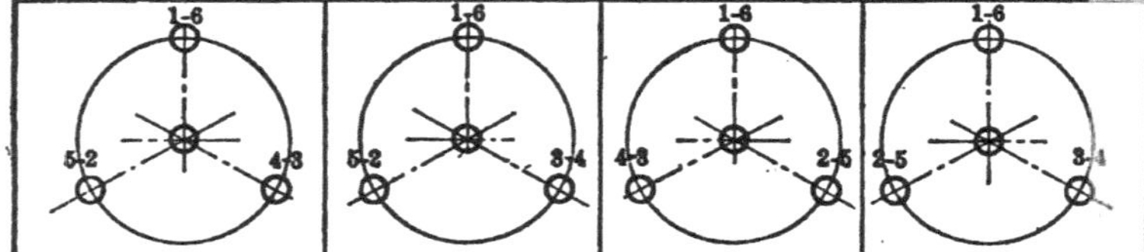

Crank Arrangement

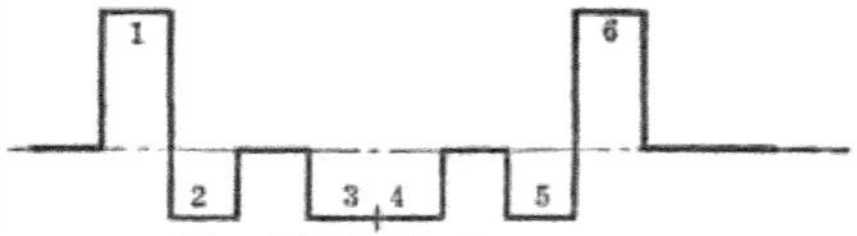

For Six Cylinders.

Positions of Pistons in Six Cylinders.

Firing Order.	Cylinder No. 1	Cylinder No. 2	Cylinder No. 3	Cylinder No. 4	Cylinder No. 5	Cylinder No. 6
1-5-4-6-2-3	Top C.L. Beginning to Fire	$\frac{1}{3}$ from Top C.L. on Compression	$\frac{1}{3}$ from Top C.L. on Compression	$\frac{1}{3}$ down from Top C.L. on Intake	$\frac{1}{3}$ down from Top C.L. on Intake	Top C.L. Ex. Closing In. Opening
1-5-3-6-2-4	"	"	$\frac{1}{3}$ down from Top C.L. on Intake	$\frac{1}{3}$ from Top C.L. on Compression	$\frac{1}{3}$ from Top C.L. on Compression	"
1-4-2-6-3-5	"	$\frac{1}{3}$ down from Top C.L. on Intake	$\frac{1}{3}$ from Top C.L. on Compression	$\frac{1}{3}$ from Top C.L. on Compression	$\frac{1}{3}$ down from Top C.L. on Intake	"
1-2-3-6-5-4	"	$\frac{1}{3}$ down from Top C.L. on Compression	$\frac{1}{3}$ down from Top C.L. on Intake	$\frac{1}{3}$ down from Top C.L. on Intake	$\frac{1}{3}$ down from Top C.L. on Compression	"

1-5-4-6-2-3

1-5-3-6-2-4

1-4-2-6-3-5

1-2-3-6-5-4

Firing Order Chart for Four and Six Cylinder Motors.

TIMING OF MULTIPLE CYLINDER MOTORS

So far we have only considered the timing of a single cylinder and have neglected the timing relations that exist between the cylinders of the modern motor. It should be understood that the timing of each cylinder is always the same on any multiple cylinder machine, the only additional point entering into the problem being the relation of the cranks and cams of one cylinder to the cams and cranks of the next. The purpose of increasing the number of cylinders beyond unity is to secure balancing, as well as to decrease the stresses and also to reduce the weight of the fly-wheel. In regard to the timing we have only to consider the effects of balance.

Every cylinder of the four-cycle type has an explosion or power impulse in every two revolutions. With two cylinders there are twice as many impulses, or an impulse in every revolution. With four cylinders the impulses are again doubled, giving two impulses per revolution. For the same power output the power given out in a two-cylinder motor per impulse is one-half of that of the single-cylinder, thus causing one-half of the shock on the machine and the passengers. It will be seen that increasing the number of cylinders for a given power decreases the shock, and that with an increasing number of impulses per revolution the "torque" or pull on the machine is made steadier and therefore more effective.

With more than four cylinders the impulses "overlap," that is to say, there is always two or more cylinders active at one time. With six or eight cylinders the torque is modified so that the effects of individual cylinder impulses are scarcely perceptible.

In addition to the reduction of individual impulses, it is possible to balance the multicylinder type mechanically without massive counter-balances on the crank shaft.

Very good mechanical balance may be had with the well known two-cylinder opposed type, since the crank is arranged in such a way that the pistons move in opposite directions at the same time, and the connecting rods form equal angles with the cylinder center lines. This type is not well balanced in regard to the explosive impulses. In the case of a four-cylinder motor, the crank shaft is arranged so that two pistons move up and two move down simultaneously. The connecting rods do not make equal angles with the cylinder center line as in the case of the two-cylinder opposed. The arrangement of the crank of a four-cylinder motor is shown by Diagram A, Fig. 6, in which the dot and dash line represents the center line of a crank shaft. The crank throws, marked 1-2-3-4, show the relative positions of the four pistons, the pistons of cylinders 1-4 being at the bottom of the stroke and 2-3 at the top of the stroke. An end view of the crank circle is shown at the left in which the small circle 2-3 represents the crank pins at the top, and 1-4 the pins at the bottom.

It is possible to fire this arrangement in two ways with evenly distributed impulses: (1) Cyls. 1-2-4-3, and (2) Cyls. 1-3-4-2, the numbers given representing the number of the cylinders, while the arrangement of the numbers gives the order in which they are to be fired. A table at the top of the chart gives the events taking place in the various cylinders with the two different firing orders. The crank pins of a four-cylinder motor are all in the same place, or all lie on the same vertical line, as shown in the end elevation. It is usual to take the front cylinder as "No. 1."

With a six-cylinder motor it is possible to fire in four orders—

> 1-5-4-6-2-3
> 1-5-3-6-2-4
> 1-4-2-6-3-5
> 1-2-3-6-5-4

It should be remembered that the crank throws of a six-cylinder motor are not all in the same plane, but are arranged at an angle of 120 degrees apart. As two cylinders move together there are always two throws in the same position, as will be seen from the four end views at the left of the six-cylinder crank arrangement.

The table at the bottom of the chart gives the events that are taking place in the different cylinders with the various firing orders. For this table we are indebted to C. T. Schaefer.

FIRING ORDER OF EIGHT CYLINDERS

In general, there are eight orders in which an eight-cylinder motor can be fired, but there are only two of these that can be used if full advantage is to be taken of the eight-cylinder principle. The orders adopted by well known manufacturers follow:

Cadillac Eight. Two blocks of four cylinders each are used on the Cadillac, the cylinder blocks being at an angle of 90 degrees with one another. The blocks fire alternately, first on the left and then on the right, it being assumed that the observer is facing the front of the motor. The numerals as arranged below each represent a cylinder, while the two columns represent the right and left blocks. To follow the firing follow the numbers in order, 1, 2, 3, etc.

5	2
3	8
7	4
1	6
Right	Left
Block	Block
Front	of	Motor

With front cylinder No. 1 on the left firing first, the next cylinder will be No. 2, the last cylinder on the right. It will be noted that if the succession on any one side is followed that each block fires in the usual four-cylinder order—1-3-4-2. The timing will be found in the timing table.

With No. 1 starting on working stroke, No. 2 is on compression, No. 3 is starting compression, No. 4 is completing suction, No. 5 is starting suction, No. 6 is scavenging, No. 7 is exhausting, and No. 8 is about one-third through working stroke.

The De Dion Eight. The blocks of the French De Dion eight also fire alternately, but in the reverse way from the Cadillac, the first cylinder starting on the right instead of the left. There will be found a considerable difference between the two in the firing order.

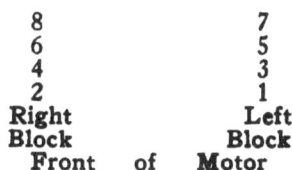

8	7
6	5
4	3
2	1
Right	**Left**
Block	**Block**
Front of	**Motor**

The order is to be followed in the same way as in the diagram given with the firing order of the Cadillac. The two systems given give two extreme firing orders possible with the eight-cylinder motor working under proper conditions.

The White Eight-Cylinder (1916). Instead of having the cylinders arranged in the conventional "V" form, the White eight-cylinder motor has a row of four vertical cylinders with the two trunk pistons in tandem, that is, a single cylinder unit is provided with two bores, one above and the other below, with a common piston the diameter of the lower piston being larger than the upper by an amount necessary for equal areas. A single piston casting serves for both cylinders, and is connected to the crank shaft by a single connecting rod to a single crank throw.

A set of exhaust and inlet valves is used for the upper and lower chambers of each cylinder unit, but are located on opposite sides of the motor. This may be considered as being two superimposed "L" head cylinders formed in one casting, the valves of which are driven by two cam shafts. All crank throws are in the same plane as in the ordinary type of four-cylinder engine. A single carburetor is used.

Starting at the front with both the upper and lower cylinders and numbering toward the rear, the cylinders in the top row will be successively, 1-2-3-4, and from the front to the back on the lower row the numbers will be 5-6-7-8. The firing order will be 1-8-2-7-5-4-6-3.

TIMING OF KNIGHT MOTOR

The valve functioning of the Knight motor is performed by two concentric sliding sleeves surrounding the piston, and not by the poppet valves ordinarily used. Instead of cams, the sleeves are driven by eccentrics from a shaft that corresponds to the cam shaft of the poppet valve motor.

Since a description of the mechanical features of the Knight motor would be out of place in this book, the reader is referred to the "Gas, Oil and Steam Engine," published by the C. C. Thompson Publishing Company, Chicago, Ill., the publishers of this volume.

Make	Inlet Opens Past Upper Center	Inlet Closes Past Lower Center	Exhaust Opens Before	Exhaust Closes After
Stearns-Knight	4°	40°	60°	Top Center
Moline-Knight	20°	50°	50°	5°
Lyons-Knight				
Porter-Knight Racer......	5°	62°	70°	15°

It will be noted that the Porter-Knight motor has the exhaust port open 10 degrees after the inlet opens, and that the Moline-Knight has the exhaust closed 15 degrees before the inlet opens.

TIMING THE GNOME AERO MOTOR

The seven-cylinder rotary Gnome motor fires alternate cylinders in succession. Starting with cylinder No. 1 on the firing point, the order will be taken against rotation:

1-3-5-7-2-4-6

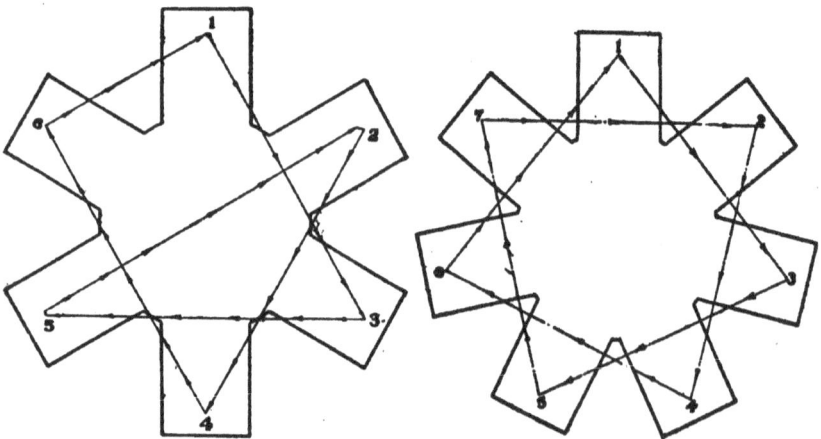

Fig. 7.—Firing Order of Gnome Motor.

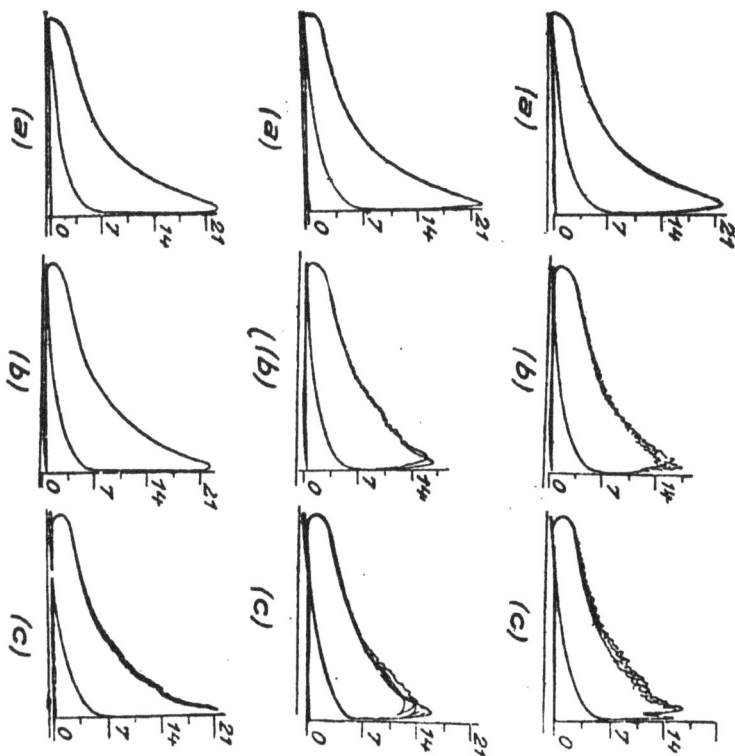

Effects of Spark Timing Shown by Indicator Cards. Left Vertical Row (a-a-a): at Top Shows Correct Timing, Center Shows Retarded Spark; Bottom Shows Advance. Center Vertical Row: (d-d-d) at Top Shows Retard and Weak Mixture; Center (d) Is Correct Spark and Weak Mixture; Bottom (d) Advanced Spark and Correct Mixture. Right Hand Vertical Column (c-c-c); Normal Spark and Weak Mixture; Advanced Spark Rich Mixture; Bottom Preignition.

TIMING OFFSET CYLINDERS

It is common practice to "offset" the crank shaft from the cylinder center line. There are two advantages to be gained by offsetting, (1) The pressure of the piston on the cylinder walls is much reduced,

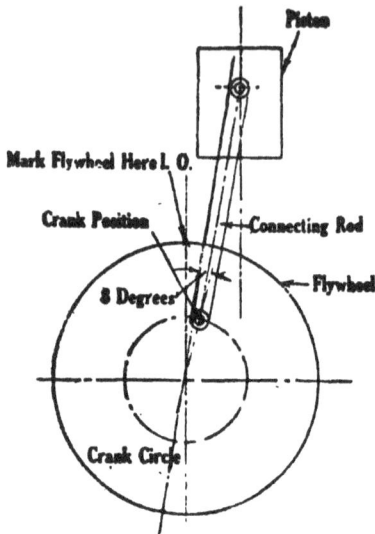

1 Inlet Opening

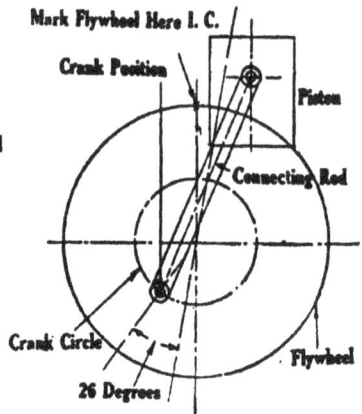

2 Inlet Closing

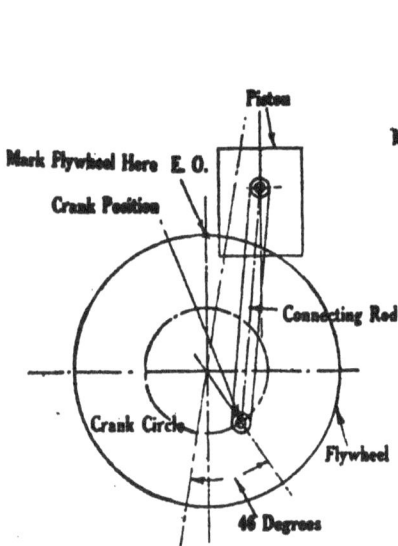

3 Exhaust Opening

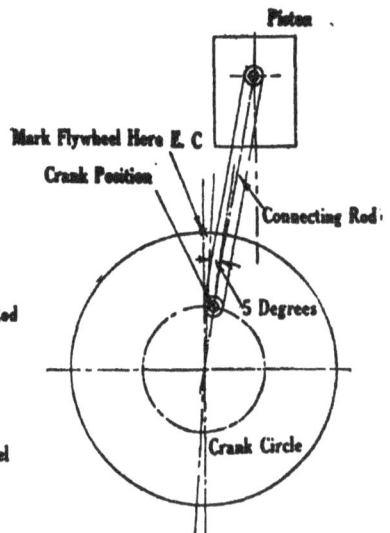

4 Exhaust Closing

Diagram for Setting the Valves of an Offset Motor.
(Courtesy "Automobile Journal.")

as the connecting rod makes a smaller angle with the cylinder. (2) The working stroke is increased and the compression stroke is shortened.

Practically the only difference in timing the valves of an offset motor is in locating the dead center. With no offset the two dead centers lie on the cylinder center line. With the offset motor, the crank pin lies to one side of the cylinder center line when on dead center, and the three points of crank shaft, crank pin and cylinder center are on one straight line. In other words, the upper point of the piston in an offset motor occurs when the crank pin is to one side of the vertical center line. Bringing such a crank to the vertical or upper cylinder center line will lower the piston.

This is illustrated by Fig. 8, which shows four crank positions for four valve events. The following timing has been assumed:

Diag. 1. Inlet opens at 8 degrees past upper dead center.

Diag. 2. Inlet closes at 26 degrees past lower dead center. Inlet opening = 198 degrees.

Diag. 3. Exhaust opens 46 degrees before lower dead center.

Diag. 4. Exhaust closes 5 degrees past upper dead center. Exhaust opening = 231 degrees.

It should be noted that this timing is only used for an example in working out the diagram and that the same methods can be applied to any other timing. The first thing to do will be to take off the lower half of the crank case so access may be had to the crank. Now by inserting a wire through the cylinder head and by resting the end on the top of the piston, the shaft may be slowly turned back and forth until the extreme top of the piston travel is found. This is the upper dead center. A much more accurate method would be to use a straight edge, so as to get the three centers of the piston pin, crank pin and crank shaft center in line, but as this is an exceedingly difficult thing to do on account of the construction of the case, we will assume that the operation is performed by the first method.

After the top center is found, lay a straight edge across the lower crank case edge and measure the angle made by the crank with the straight edge. This will be the angle of offset or the angle made by the crank with the cylinder center line when the piston is on the upper dead center.

Add to this the angle of inlet valve advance (8 degrees) and turn crank to new angle. This will be the crank position at which the inlet valve should start to open. Find the following lower dead center and in the same way locate the point at which the inlet valve closes. The exhaust valve positions may now be found in the same way.

Care should be taken to mark each of the positions on the fly-wheel when located. A pointed or indicating finger should be placed on some convenient part of the engine (near the fly-wheel) so that the marks on the wheel can be relocated. We are indebted to the **"Automobile Journal"** for the diagrams illustrating this article.

VALVE SETTING ON STATIONARY ENGINES

The exhaust should open when the crank lacks 30 degrees of completing the outer end of the power stroke, that is, the crank should make an angle of 30 degrees with the center line of the cylinder when the exhaust valve begins to open, and should be inclined away from the cylinder. Some makers have the exhaust open a little later in the stroke, but little is to be gained with a later opening, as the retention of the charge beyond 30 degrees heats the cylinder and does very little towards developing power. The only advantage of the late opening is that the valve opens against a lower pressure and causes slightly less wear on the parts.

The exhaust valve should close 5 degrees after the crank has passed the inner dead center on the exhaust or scavenging stroke, although some makers close the valve exactly on the dead center. The 5 degrees should be given to allow the gas all possible chance of escape. The piston is said to be on the inner dead center when it is in the cylinder as far as it will go, and on the outer dead center when it is on the center nearest the crank shaft.

The intake valve should open about 5 degrees after the exhaust valve closes, or 10 degrees after the crank passes the inner dead center. The inlet valve should never open before the exhaust valve closes on a low speed engine. The above timing is for engines running 150-600 R. P. M. The automatic type of inlet valve, of course, cannot be timed, but attention should be paid to the strength and tension of the spring and the condition of the valve stem guides.

The inlet valve should close 10 degrees after the crank passes the outer dead center in order that the cylinder be filled to the fullest possible extent. If the valve closed exactly on the dead center, a partial vacuum will exist and the charge retained in the cylinder will be comparatively small, but if the valve remains open past this point the air would have time to completely fill the cylinder and develop the capacity of the engine. The longer the inlet pipe, the longer the inlet valve opening.

PART XI

FORD IGNITION TROUBLES

The Ford magneto is of the flywheel type, a series of horseshoe magnets on the flywheel running past a corresponding series of coils mounted on a stationary support. As the poles of the magnets are mounted well out toward the periphery of the flywheel, they pass the coils at a very high velocity, even when the wheel is revolving at a low number of revolutions per minute. This allows the generation of a considerable current at low-motor speeds. Since the coils are stationary, there are no brushes or other forms of revolving contacts in the magneto proper. The current thus generated is alternating current of low voltage, and is stepped up to the high-tension current by means of vibrator coils of a type already described. A timer, or commutator, is used that is similar to a battery timer in every respect, this device being attached to the front end of the camshaft.

Fig. 1 shows the arrangement of the magneto in detail, the flywheel magnets at the left and the stationary coil plate at the right. The magnets M, sixteen in number, are attached to the flywheel A by the bolts B. The outer ends or poles of the magnets are fastened by the magnet clamps C. As will be seen from the left-hand diagram, similar poles of adjacent magnets are joined together under each magnet clamp, two north and two south poles alternately. As the different polarities pass a given pole, the generated current alternates in direction with the change in the polarity of the magnetic flux. The current thus produced cannot be used for storage-battery charg-

209

ing unless passed through some type of rectifier. If the current is to be used for lighting it can be obtained only when the engine is running.

As shown by the right-hand diagram, there are sixteen magneto coils that correspond in spacing and radius to the poles of the magnets. As installed in the machine, the coils face the magnets instead of being in the position shown, the illustration being reversed for clearness of explanation. Each coil consists of a number of turns of copper wire or strap, thoroughly wrapped with insulating tape, and is mounted on a central iron core. In the figure the coils are indicated by R and the cores by S, the supporting plate being marked U. The pole tips of the magnets pass over the faces of the cores S with about 1/32-inch clearance betwen the faces of the cores and the pole tips. All of the coils are connected in series, one end grounded to the plate, and the other end is connected to the connection post P, which is mounted on the transmission housing. From the connection post wires run to the spark coils and lamp circuit. Connection between the terminal of the coils and the connection post is had by a contact spring.

Ford Magneto Troubles and Adjustments. In the first place, the reader must be cautioned against connecting a storage battery in circuit with the magneto unless a double-throw switch is used, which will disconnect the magneto when the battery is in service. If battery current passes through the magneto coils, it is almost certain that the magnets will become demagnetized and worthless. In case of such an accident the magnets can be remagnetized by the owner of the car, but the best and cheapest method is to replace them with new magnets. These can be obtained from a Ford branch house or service station. For the benefit of those that wish to remagnetize the magnets we will give the following instructions:

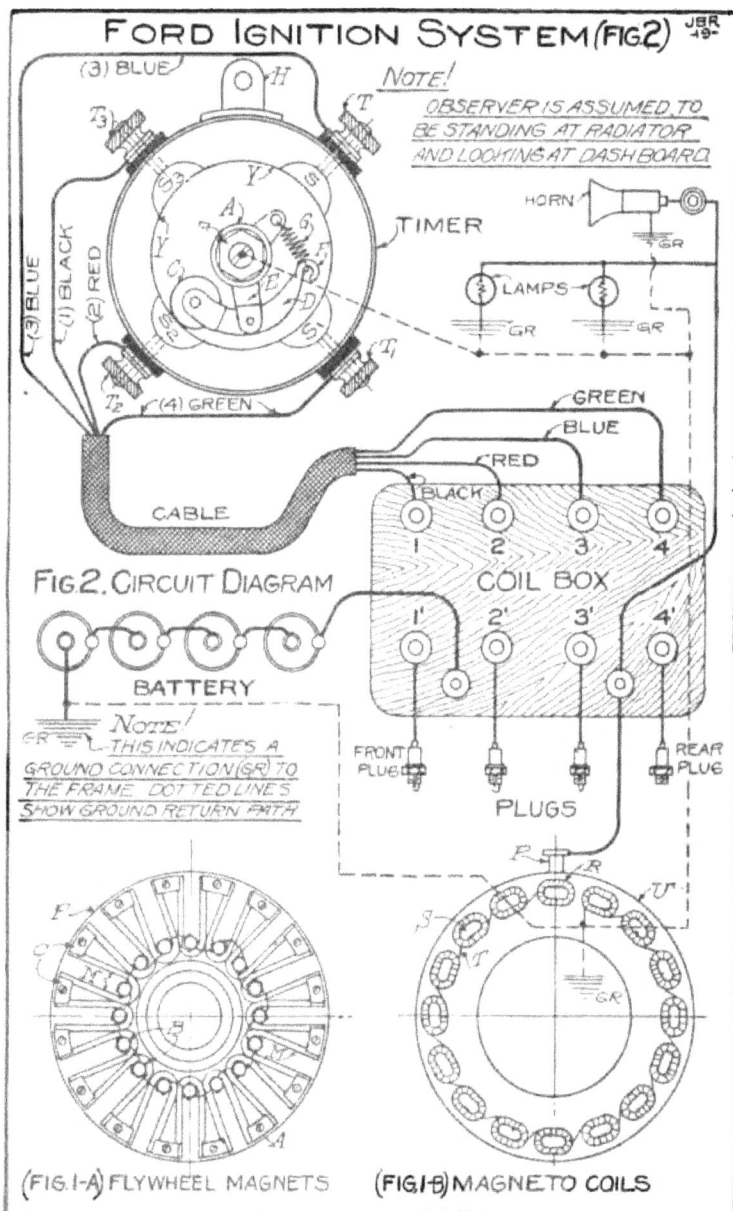

FORD IGNITION SYSTEM (FIG2)

(3) BLUE

NOTE!
OBSERVER IS ASSUMED TO
BE STANDING AT RADIATOR
AND LOOKING AT DASHBOARD

TIMER

HORN

LAMPS

(3) BLUE   (1) BLACK   (2) RED

(4) GREEN

GREEN
BLUE
RED
BLACK

CABLE

FIG 2. CIRCUIT DIAGRAM   COIL BOX

BATTERY

NOTE!
THIS INDICATES A
GROUND CONNECTION (GR) TO
THE FRAME. DOTTED LINES
SHOW GROUND RETURN PATH

FRONT PLUG   REAR PLUG

PLUGS

(FIG. I-A) FLYWHEEL MAGNETS   (FIG. I-B) MAGNETO COILS

The process of remagnetization requires a direct current of at least 30 amperes. This can be obtained from direct-current lighting mains, or better, from four to six storage batteries connected in series. Cells having a capacity of 40 amperes at 6 volts will be suitable. The magnets are turned until they are in correct relation with the magneto coils, and the current is then passed through the coils at short intervals until the magnets are restored to their former strength. Each coil acts as an electromagnet and passes its magnetic flux through the permanent horseshoe magnet, a portion of this flux being retained. If lighting current is used, some form of resistance must be placed in the circuit or the high voltage will most certainly burn out the coils, leaving the machine in a worse fix than ever. In this regard be sure that the lighting current is direct and not alternating, for if alternating current is passed through the coils the magnets will be entirely demagnetized.

After obtaining a source of current, place an ordinary hunting compass on the transmission and over the flywheel, so that the compass is ⅞ inch to the left of the connection post. The compass, we will assume, has the North-pointing end of the needle colored blue, with the South-pointing end left the natural color of the steel. In placing the compass to the left of the connection post, we are supposed to be facing the front of the car. Turn the motor over slowly until the blue hand of the compass points to a spot about ¼ inch from the side of the fiber bushing at the bottom of the connection post. Disconnect the wire from the connection post and attach instead the positive (†) wire of the battery. Now, take the negative battery wire and touch it quickly and intermittently to the frame or some other metal part of the car, leaving the wire in contact with the frame for about 5 or 6 seconds. As the wire is removed from contact with the frame a bright flash will be seen if the circuit is in condition.

About twenty such contacts should be made at intervals of one second before the charging is considered as complete. Be sure that the positive battery wire is connected to the post and that the blue end of the compass needle points to the connection post. This is rough on the battery because of the large amount of current drawn, and the contacts should not be made for a longer period than absolutely necessary. Better buy new magnets.

A gradually weakening magneto current may be due either to weakened magnets or to dirt accumulating under the contact spring that completes the circuit between the magneto coils and the connection post on the transmission cover. It may also be caused by short-circuited or grounded coils, although this trouble is rather unusual. Try the simplest remedy first—that is, remove the three screws that hold the binding post in position, remove the spring, and then clean out any foreign matter that may have collected at the point of contact. If this does not remedy matters, carefully examine the wiring for short circuits or grounds before attempting to undergo the tortures of removing the magnets from the magneto. Many times the wear in the crankshaft bearings will cause the magnet and coils to separate to such a degree that the current will be weak and the engine will misfire at moderate speeds or be difficult to start. Trouble from this source will be saved if the gap between the coils and magnets is measured occasionally. It should be little different than 1/32 inch. If the separation is too great the current will be weak and the magnets will lose life rapidly. If the gap is less than 1/32 inch there is likelihood of the magnets rubbing on the cores of the cells. Rubbing often results in the destruction of the coil insulation, causing grounds and a sudden loss of brightness in the lamps.

When the magneto stops business suddenly, while at the same time the motor runs well on the battery, clean out the contact spring under the connection post, or see that the wire and coil insulation is not abraded through rubbing. Magnets do not weaken suddenly. If very large lamps are used, or too many are installed, the demand for current will be greater than the supply, and the trouble will usually be blamed on the magneto. If there is trouble with the ignition after the installation of new lamps, this is probably the cause. The lamps furnished by the Ford company consume 2 amperes at 8 volts, and the best results will be obtained by lamps taking this voltage and current. The many current regulators now built for the Ford magneto improve the lighting wonderfully and save current for the ignition system.

Should the magnets be found at fault it will be necessary to remove the entire magneto—that is, it will be necessary to remove the entire power plant from the car before beginning the real work. If you are wise you will try everything connected with the ignition system before starting this job. After the power plant is removed, take off the crankcase and transmission cover, and unfasten the four bolts that hold the flywheel to the crankshaft. Pull off the flywheel and the entire magneto is exposed. All of these parts should be marked so that they can be replaced in their proper positions.

To take off the old magnets, remove the cap screw and the bronze screw that hold them in place on the wheel, carefully noting the method of attachment. The new magnets as received from the Ford service station are mounted on a board in exactly the same relative position as they will occupy on the fly-wheel. Be careful that they are placed on the wheel in the same position as on the board, as a slight mistake will neutralize the polarity of one or two pairs of magnets and hence cause unlimited trouble with the ignition. Should such a mistake occur remember that poles of like polarity should lie together under the same magnet clamp. The second important item is that of clearance between the magnets and coils.

Line up the faces of the magnets so that they are just 1/32 inch from the faces of the coils spools or cores. While the magneto is in this disassembled condition take a look at the coils—see whether the insulation is worn or whether the connections are loose, paying particular attention to the ground connection made by the copper ribbon with the coil frame. If the ground connection is loose or broken, no current can flow since the return to the magneto is through the frame or ground. The circuit must be complete at all points.

The Ford Commutator and Its Adjustment. The magneto current (Low tension) is distributed to the four spark coils by a roller type commutator or timer. This instrument determines the time at which the spark occurs in the cylinder and distributes the current to the coil of the cylinder next in firing order. In effect it is a rotary automatic switch that revolves in a fixed relation to the crank-shaft, its function being to make and break the current in the primary circuit. Since there are four cylinders, there are also four insulated contact points with which the roller makes contact as it revolves. From each contact point, is a wire lead to the corresponding spark coil. The commutator is located at the front of the motor and is driven directly from the front end of the cam-shaft so that it revolves at one-half crankshaft speed.

Fig. 2 shows a detail view of the Ford commutator and its connection to the spark coil and magneto. The roller C rotates in a hinged fork D, and is pivoted to the cam-shaft A by the pin E, the latter being fastened to a lug projecting from the cam-shaft. A small coil spring G connected with the forked lever at F forces the roller into contact with the contact segments S-S1-S2-S3 as it revolves. The contact segments are imbedded in insulating material so that they cannot make contact either with each other or with the frame of the commutator. The roller of course is grounded to the cam-shaft, so that when the roller makes contact with any segment, that segment and the connecting wires are also grounded. This arrangement allows the current from the coil to flow along the wire, through the corresponding segment and back to the magneto through the ground. The contact segments S-S1-S2-S3 are provided with the binding posts or connection posts marked respectively T-T1-T2-T3 to which the spark coil wires are attached. A short lever H attached to the commutator housing allows it to be tilted back and forth for retarding and advancing the spark.

To facilitate the connections made with the coil, the wires are colored and numbered to correspond with the cylinders and coils. Thus the blue wire No. 3 connects with the terminal T on the commutator and with the primary binding post (Upper row) on the spark coil. With colored insulation the wires can be recognized with certainty at either end of the cable. The colors on the diagram correspond with the colors of the insulation on the machine, and they should not be connected in any other order than that given. In the diagram we are supposed to be facing the engine from the front of the car, looking toward the dash.

Looking at the spark coil box we notice that there are two principal rows of binding posts, an upper and a lower row, each row having four posts. The upper row connects with the primary wires coming from the commutator while the lower row contains the high tension posts that connect with the spark plugs in the cylinders. In the lower row, post No. 1 connects with cylinder No. 1, this being the front cylinder in the block. The remaining cylinders are numbered in order, so that cylinder No. 4 is the last or rear cylinder in the block. Below the high tension posts are two low tension posts that connect with the magneto, battery, and lamp circuits. The battery post is at the left, while the post at the right connects both with the magneto and with the horn. These connections are controlled by a switch in manner that will be described later. The electric horn is also connected with the right hand connection post.

If the commutator is to perform its functions properly the roller and the contact points must be kept clean and smooth. The inside surface of the circle (Y) around which the roller runs should be perfectly smooth so that it will make perfect contact and not skip and bound over the obstructions that can be caused by foreign matter or worn places in the insulation. If the roller fails to make good contact, the cylinder corresponding to that contact segment will not fire. If dirty clean thoroughly with a cloth moistened in gasoline. If the parts are worn replace them with new or reface the insulation and segments. The spring should be strong enough to force the roller into good contact. Sometimes the roller spring weakens and causes a bad case of missing. Worn insulation on the wires leading from the commutator is a common cause of trouble as in this case they produce short circuits. The back and forth movement of the commutator housing in advancing and retarding the spark often causes broken wires or loose connections. Always see that the wire connections are tight and clean. A steady buzzing in one of the coil units will indicate a short circuited wire. A short-circuited commutator wire is likely to cause a severe "kick-back" when the car is cranked for the short is likely to close the circuit before the pistons reach the end of the compression stroke.

- The commutator should be sparingly oiled, enough to prevent friction and wear but not enough to gum up the contacts and pivots. In cold weather the oil often thickens to such an extent as to prevent the roller from making contact with the contact segments. This is often the cause of cold weather starting trouble as the roller is often unable to scrape the oil away from the segments.

Even after getting the motor started, one or more cylinders will continue to misfire until the oil is sufficiently softened by the heat to permit of good contact with the rest of the segments. A little kerosene mixed with the lubricating oil will prevent the oil from solidifying in extremely cold weather.

For repair or adjustment, the commutator can be removed by removing the spark control rod from the timing lever H, and then loosening the screw that passes through the breather tube on top of the cam-shaft gear cover. This will release the commutator case so that it can be easily removed for inspection. The brush can be removed from the cam-shaft by unscrewing the lock nut removing brush cap and driving out the pin. When the brush is replaced it must point upward when the exhaust valve of cylinder No. 1 is closed.

Oil the commutator slightly every day, a few drops at a time being sufficient. The roller revolves at a high speed and if not properly lubricated will soon wear out and cause missing. At the end of a week, the spent and carbonized oil should be removed or there will be trouble due to sticking of the parts through the thickening of the oil. Thin oil should be used or heavier oils should be thinned with kerosene, the thinning of the oil being of more importance in the winter than during hot weather. The parts are small and easily gummed up with heavy lubricants such as cylinder oil. Any trouble due to rough or dirty surfaces is more noticeable at high speed than at low as the roller starts jumping over the high spots at anything greater than a very moderate speed.

The switch has three positions. One for connecting the battery into circuit (Bat.); one position for connecting the magneto (Mag.), and a central position (Off) for cutting out or stopping the engine.

PART XII

BOSCH MAGNETO CHART

A chart developed by the Bosch Magneto Company obviates many of the troubles experienced in timing a magneto. As the majority of magneto builders specify the opening of the circuit breaker to take place at some particular crank angle (in degrees) it is often difficult to find the angle because of the variation in piston travel and connecting rod length. With the automobile it is impossible to make measurements directly because all of the ports are enclosed.

As the distortion due to the length of connecting rod is practically the same in all cars, the calculation for this factor may be neglected, except in extreme cases, the ratio of the crank to rod being taken as 1 to 4.5 in all cases.

The row of figures at the bottom of the chart gives the extreme stroke of the piston or twice the length of the effective crank arm. The vertical row of figures at the right are an index to the slanting lines and give the angle made by the crank with the center line of the cylinders in degrees. The figures at the left index the horizontal lines and designate the travel made by the piston for a given crank angle.

Example. We will assume that our engine has a stroke of six inches, and that it is necessary to find the distance traveled by the piston from dead center when the crank has moved through an angle of 30 degrees with the cylinder center. In other words, we wish to

obtain the crank position for a magneto whose breaker is to open
the circuit at 30 degrees past upper dead center.

Locate the given stroke in the figures at the bottom of the chart

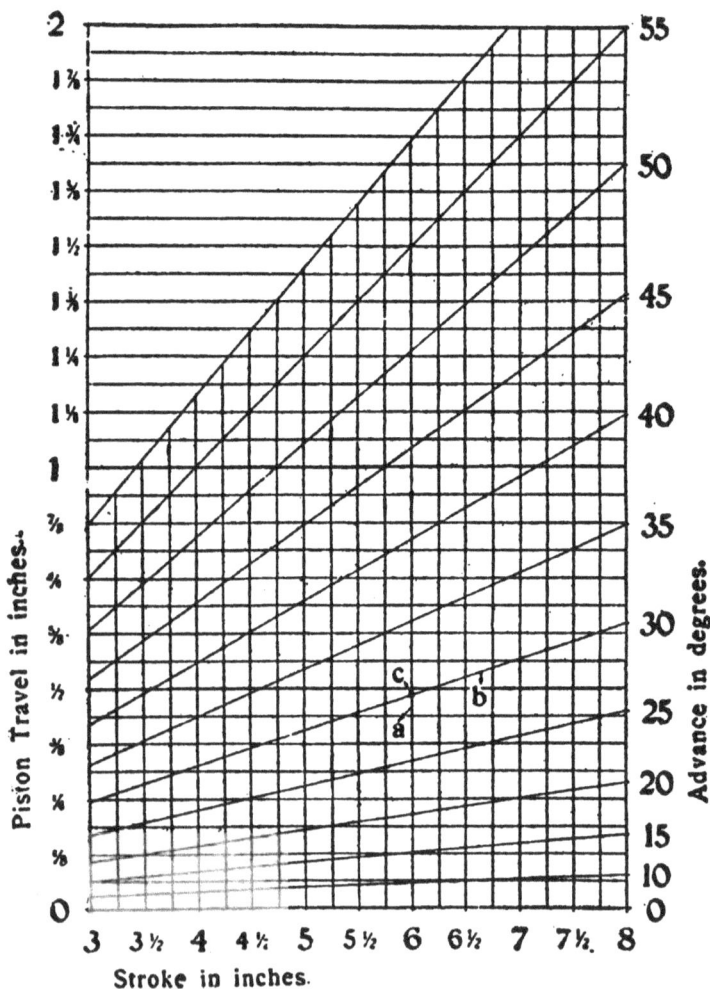

Fig. 1.—Bosch Magneto Timing Chart.

and trace upwardly along the vertical line until it intersects the slanting line indicated by "30" at the left, as at the point "C." Follow
the nearest horizontal line to the left hand edge of the chart where
it will be found to nearly coincide with the ½-inch numeral. This

means that for an angle of 30 degrees, the piston must travel ½ inch down from the upper dead center.

Now, to set the engine at 30 degrees, set the engine on dead center, insert a wire through the petcock in the cylinder head, and then turn the engine over slowly until the piston has moved down ½ inch, as shown by the wire. The crank is now at 30 degrees.

This chart is equally applicable to setting valves or in adjusting any part that requires setting of the crank at any specified angle.

CHARGING MAGNETS

When the permanent magnets become so weakened that the magneto will no longer give a good spark, they must be removed from the frame and recharged. This is best done by placing them on the poles of a pair of powerful electric magnets as shown by Fig. 2 in which A is the permanent magneto magnet, and M-M¹ are the electromagnets. The current from the line enters the magnet coils as at L and leaves at L¹, highly magnetizing the pole pieces N and S. The flux from the poles passes through the permanent magnet A, practically saturating the steel.

It should be remembered that only **direct** current can be used with this device, since an alternating current would reverse so rapidly that the magnet A would be in worse shape than before. There are several remagnetizers on the market that utilize the direct current from the ignition or lighting batteries. In the figure, Y is the soft iron yoke that fastens the magnets M and M¹ together and also completes the magnetic circuit.

When demounting the magnets for remagnetizing their ends should be marked so that they will reassemble correctly on the magneto. The marked magnet legs should be placed on the same pole of the electro-magnets so that the legs on one side of the magneto armature will all be of the same polarity. It is a good plan to place all of the right hand legs on north or right hand (as shown in figure) poles of the electro magnet. Failure to observe this precaution will mean trouble with the magneto, since placing dissimilar legs together is equivalent to a magnetic short circuit.

After placing the permanent magnets on the poles S-N, close the switch and start tapping the magnet A rapidly and lightly with a wooden mallet or soft faced hammer. Hammering the magneto magnets while under the influence of the electros seems to increase the magnetic flux and makes the maximum flux more permanent. Under no conditions hammer the magnets when the current is off or when they are removed from the charging apparatus, as this seems to literally knock the magnetism out of them. Leave the current on for at least two minutes.

Before shutting off the current place a soft iron bar or "keeper" B across the poles of the magnets as indicated by the dotted lines and keep this bar on until the magnets are well shoved down over the pole pieces of the magneto. Without the keeper, permanent magnets lose their strength in a very short time after the magnetizing current is shut off. Never allow magnets to lay around after dissembling without placing two or three keepers across their poles.

Fig. 3 shows a method of charging magnets without the use of the apparatus just described. It is a tedious and crude method, but often is the only means that the layman has of performing the work.

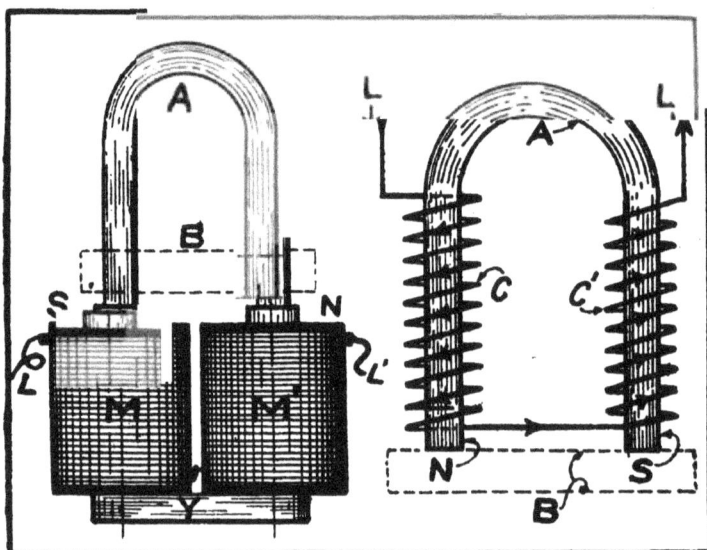

Fig. 2.—Charging Magnet.   Fig. 3.—Temporary Rig.

Two coils of wire, C and C1, are wrapped around the legs of the magnet A in **opposite** directions. The wire should be comparatively heavy, say No. 18 gauge, and there should be at least 75 turns on each leg. After completing the winding a keeper B should be placed across the poles N and S. The ends L and L1 should be connected to the source of **direct** current.

Tap the magets as before while the current is passing, and at the end of two minutes rapidly slip off the coils, replace the keeper, and mount on magneto. If electric light current is used from 110 volt main it will be necessary to introduce some form of resistance into the circuit to prevent too sudden a rush of current. About 10 amperes will usually be required, and even more if you can obtain it. With a storage battery, the coils must not be left in circuit more than 5 seconds, for the battery is likely to be damaged through the heavy discharge of current due to the low resistance of the coils.

Do not forget to wind the coils in **opposite** directions, nor to use only **direct** current. The wire should be copper, preferably insulated with a double cotton cover.

Before starting the charging process with the first method, in cases where you do not know the character of the current, turn on the current and place the keeper across the poles of the magnets. If a humming noise is heard do not attempt to charge, as the humming is always due to an alternating current.

TIMING MAGNETOS

In any gas or gasoline engine, combustion must be complete at the end of the compression stroke, or as the crank pin reaches the upper dead center. As all mixtures require a certain length of time to complete combustion after the application of the spark it is evident that the spark must occur slightly before the end of the compression stroke. This "advance" of spark, numerically, depends upon the quality of the mixture, upon the compression pressure, and upon the speed of the engine. Poor mixtures require more time for combustion than normal mixtures, while high compressions reduce the volume through which the flame must spread and therefore reduce the advance. High speeds require more advance than low since there is less time in which to burn the mixture and consequently the combustion must be started earlier.

From the three variable quantities it will be seen that the magneto must be capable of adjustment to meet all running conditions or so that the sparks can be made to occur over a wide range of piston positions. A fourth factor of the timing range is that of engine construction for, with all other items equal, the advance and retard are affected by the size and shape of the combustion chamber and the position of the spark plugs.

Unlike the case with the common battery system, the magneto of the true high tension or transformer has no vibrator lag, and little if any, electrical inertia. For this reason the advance and retard angle of the magneto is less than with the ordinary battery equipment. Since all cylinders are fired by the same magneto circuit breaker, every cylinder receives the spark at the same point in the stroke. With batteries the latter condition is only approached with the master vibrator system.

In general the magneto is timed or geared to the engine in such a position that the spark occurs on, or a trifle before, the end of the compression stroke with the timing lever in the fully **retarded position.**

It should be noted here that in all cases measurement is taken with the magneto fully retarded. With some engines, notably those of the "T" head type, the spark is made to occur at full retard at a point ¼ inch before the piston reaches the end of the stroke. The exact point is determined to some extent by the valve setting or construction of the combustion chamber. Some engines require considerable advance while others will stand none, or very little. The best point on engines not already equipped for magnetos is found by trial.

In the case of timing magnetos on engines with marked fly-wheels the timing is a simple matter, for then one only has to place the pointer opposite to the marks on the fly-wheel and set the breaker to open at this point with full retard. For simplicity in connecting up the plugs, start with cylinder No. 1 in the firing position as indicated on the fly-wheel.

Before continuing further on the subject of timing it should be understood that the spark occurs at the point where the circuit breaker contacts are just barely beginning to open. The spark occurs when the contact is broken by less than 0.005 inch, so that the points must be observed very carefully or measurements must be taken from the armature or from points marked by the builder of the magneto. A small error in this observation will make a very considerable error in the timing. Some makers provide a key, which, when inserted into the armature through the frame, hold the breaker at the point where it just starts to open. This is very convenient as it requires no particular care on part of the person doing the timing. With some magnetos, a gauge is provided, while with others it is necessary to measure the distance between the edge of the armature shuttle and the edge of one of the pole pieces. Owing to these differences it will be necessary to specify procedure with several of the principal makes of magnetos.

The first step in timing will be to set Cylinder No. 1 on the top center (or in firing position if marked on fly-wheel). In some motors the firing point "No. 1   ," is found near the top center "No. 1. T. C.," while in others the two points coincide on the mark "No 1. T. C." Look carefully for the firing point and if found set this opposite to the pointer instead of the T. C. mark.

When the wheel is not marked we must set the engine by directly measuring the piston position through the cylinder head. Access to the piston in most cases can be had through the relief cocks or plugs in the cylinder head or in the cases of two part cylinders by removing the head entirely. A bicycle spoke, knitting needle, or small rod can be dropped through the plug or relief cock until the end rests on top of the piston. The engine can then be slowly turned over on the compression stroke until a point is found where further turning in either direction causes the rod to move up or down. The highest point recorded by the rod is the upper dead center. (See Page 215.) This measurement is crude, impossible for use in correctly setting valves, but close enough for ignition. Should the engine be of the "offset" type accurate setting may be performed by the method shown on Page 205. But even with the offset engine, the rod measurement is close enough for timing the ignition. Fly-wheel indexing is described on Page 217 for battery ignition by Fig. 1 and will give an idea as to the procedure with a magneto setting. When once the center is found it should be permanently marked on the fly-wheel for future reference. Find and mark the bottom center in the same way.

Not only must Cylinder No. 1 be set on dead center but it must also be placed on the compression stroke, for as the complete cycle is performed in two revolutions it is an easy matter to cause the ignition spark to occur at the end of the exhaust stroke which is also on the upper dead center. Should this error be made the engine of course would refuse to start.

The compression stroke can be identified by the order in which the valves of a given cylinder open and close. Turn the engine over slowly by hand in the direction in which it rotates when in operation and carefully watch the movements of the inlet and exhaust valves of Cylinder No. 1. Finally a point will be found where the exhaust closes and the inlet opens, together or nearly together. This is the beginning of the suction stroke. On turning a little more than one-half revolution, the inlet valve will be observed to close. This is the beginning of the compression stroke. Now keep on turning in the same direction until the top center mark already located on the fly-wheel comes opposite to the pointer. This is the upper end of the compression stroke of Cylinder No. 1. Mark on wheel (No. 1 T. C.).

Of course the approximate end of the compression stroke can first be obtained in this way, and then the exact dead center can be found afterwards. This is a matter of individual habit. In any case mark the results on the wheel before going further. Block the wheel in the dead center position so that it cannot move and mount the magneto in place with one of the magneto couplings loose on the shaft.

If you are so fortunate as to possess a magneto with a timing key for holding the armature and breaker in the firing position, turn the armature over until the key slips into position before mounting magneto on the motor bed. This holds the breaker just open in the firing position, and with the engine on dead center, it simply is a matter of bolting the magneto on the bed and fastening or keying the loose magneto coupling tight to the shaft. With this type of magneto, the timing is now completed, at least for trial. The Eisemann Type EMA and the Unterberg-Helmle (U and H) are magnetos having this feature. Remove the key from the armature and turn engine slowly over by hand to see that nothing sticks or binds. With the Eisemann, the armature shaft should be turned over until the number "1" appears at the peephole on the distributer plate top before inserting the timing key or before mounting the instrument.

Before setting any magneto see that the timing lever is set in the fully retarded position. With the Eisemann Type EMA magneto, the lead from the distributer connection No. 1 can now be connected with the spark plug in cylinder No. 1, and the remaining plug wires connected up according to the firing order of the motor. The order in which different numbers of cylinders fire can be found from the chapter on Valve Setting, Part X, pages 201 to 208. The numbers appearing in the peepholes are **not** the cylinder numbers to which the next wire is to be connected but only indicate which distributer terminal is active at that time. The terminal numbers give the successive order in which the brush makes connection with the terminals. See Fig. 3 on Page 99, and Fig. 10-A on Page 110. (The plugs are not in proper firing order as shown, but show the order in which they are connected by brush.)

Example. Say that we have a four cylinder motor in which the cylinders fire in the order, 1-3-4-2, and that we now have Cylinder No. 1 connected with the magneto to terminal No. 1. As the distributer brush connects with the terminals in their numerical order, the next live terminal will be No. 2. Since the next active cylinder, after Cylinder No. 1, will be Cylinder No. 3 according to the given firing order, it is evident that a wire from terminal No. 2 should connect with plug in Cylinder No. 3, Cylinder No. 4 will be connected to terminal No. 3, and Cylinder No. 2 will be connected to terminal No. 4.

For firing order of six, eight and twelve cylinders see pages 201 to end of chapter on valve timing, Part X.

In the above description we have assumed that the fully retarded spark is to occur exactly on dead center and that the magneto is provided with a timing or setting key. In practice, we often have different conditions to meet, for we often fire before dead center and have a different method of setting the magneto circuit breaker. The notes on the distributer connections, however, hold good for any magneto with the same number of cylinders.

Assuming for the present, that we are still firing on dead center, we will consider the method adopted on the Bosch magnetos, Types "D," "DU-4," and "DU-6," for finding the point of breaker opening. For the details of construction see longitudinal section on Page 101 on which will also be found, the part numbers referred to.

The connecting bridge (12) and the dust cap (21) are removed to determine the armature position. Fasten the magneto firmly on the motor bed, and with the driving pinion or coupling loose on the magneto shaft. The engine is supposed to be on dead center. Now turn the armature shaft slowly by hand and look through the opening left by the removal of the dust cap (21). As the armature rotates, an iron surface, and a black coil appear into view alternately, running past the pole piece. The circuit breaker will open **after** the iron edge of the armature has left the edge of the pole piece by the amounts given in the following table with right hand rotation (when facing shaft end of magneto), the gap between the armature and pole will be at the left. With left hand rotation the gap will be at the right. The length of the gap should be—

For 3 Cylinder Engine..11 to 13 millimeters..0.4331 in. to 0.5118 in.
For 4 Cylinder Engine..14 to 17 millimeters..0.5512 in. to 0.6693 in.
For 6 Cylinder Engine..21 to 27 millimeters..0.8268 in. to 1.0630 in.

When the gap is measured as above, for the required number of cylinders, hold the armature in place and firmly pin or key the loose coupling or pinion to the shaft. This is now in the correct firing position. Replace parts (12) and (21), and connect up leads to plugs as before described, taking the leads to the plugs in the order that the brush makes connection with the distributer terminals (at the magneto end), and connecting the motor ends of leads in the firing order. With Cylinder No. 1 in firing position, run lead from Plug No. 1 to the terminal connecting with that distributer segment on which the brush is resting. Determine the distributer rotation and find the next sector energized. A wire from this sector will lead to the cylinder next in firing order, and so on.

In any other magneto, the spark must be timed in the magneto by very closely observing the point at which the breaker contacts open. When, as in the majority of cases, the breaker lever is actuated by a cam striking a flat or roller follower, note particularly the point at which the cam touches the follower, and neglect the contacts for the time being. Turn the armature back and forth until the cam barely touches the follower plate or roller on the contact arm. An extremely light pressure further applied will start to separate the points. This is the point at which the spark will occur.

Usually this is far more difficult to perform than to describe since the pull of the magnets at this point tend to suddenly "flip" the armature over, and past the breaking point. With some magnetos it is almost impossible to hold the armature shaft against the pull of the magnets when holding the shaft with the fingers. After you have tried this job several times you will appreciate the genius of the man that devised the setting key. Turning the shaft by means of the coupling make the matter easier for the reason that there is more leverage, but in some cases the construction prevents this from being done.

When the spark is set to occur before the piston reaches center, the engine is first centered and is then turned back against rotation until the measuring rod moves down the required amount. Block the motor at this point, call this your center, and proceed setting the magneto as before. When the engine is not marked it is best to set the spark on the true dead center, first, for experiment. If this does not give good results then set the spark to occur before center. The Eisemann Company have a special coupling in which the two halves are divided into degrees. When the magneto is once set and is not satisfactory, the halves may be moved on one another according to the scale until the desired results are obtained without dismounting the magneto.

When a specific timing is given, by the maker of the motor, and the wheel is not marked, it is generally given in degrees measured on the crank-circle. This is an awkward measurement to make for the average man, but by aid of the magneto chart on pages 225-227 it is easily performed, since by this means the degrees can be converted into inches of piston travel and measured through the cylinder head by the rod previously described. With the stroke known, start at the bottom of the chart and follow up to the diagonal line that gives the required degrees. Follow horizontally to the left-hand column of figures which will give the distance of the piston below the top center.

Example. A motor has a stroke of 5 inches, and the fully retarded spark is to occur 25 degrees before top center. Find distance of piston below top center in inches.

Start at 5 inches at bottom of chart and trace up to the intersection of this line with the diagonal 25. Now follow horizontally to the left from this point to the left vertical row of figures where it will be found that the distance is 9/32 inch. After setting engine on top dead center, insert rod through head and turn crank against rotation until the rod sinks 9/32 inch. The crank is now at an advance angle of 25 degrees.

The timing of the magneto given in this article assumes that the breaker contacts are new or at least in good order. When thoroughly cleaned and in good order, the points should not be apart by more than 1/64 inch when fully open. If the gap is much wider than this it will interfere with the timing. Some makes, notably the Bosch, have a gauge for setting this distance. Due to burning and wear, this gap gradually increases, so that to maintain correct timing, the width of the gap between the contacts should be occasionally inspected. In the Remy transformer type the gap should be from 0.025 inch to 0.03 inch. If motor misses when idling or pulling light the gap should increased. If it misses when pulling heavy loads, decrease the gap.

PART XIII

STORAGE BATTERIES AND STARTING REPAIRS

CARE OF STORAGE BATTERIES.

Modern storage cells do not require a great deal of attention, but like any other part of the car do require occasional inspection and considerate treatment. When trouble is experienced with the circuit, and you are sure that the battery is at fault, proceed carefully for it is an easy matter to ruin the cells through improper treatment. The contents of the following table should be carefully observed in making any corrections to a lead and sulphuric acid type battery.

Discharged Cells. Never allow the voltage to drop below 1.8 volts per cell, for below this point there is danger of sulphation. It is far better to start recharging when the voltage drops to 1.9. The hydrometer test gives even a better idea of the condition of the battery than the voltmeter since it takes the condition of the electrolyte into consideration.

When all of the cells are in good condition their gravity will test within 25 degrees or "Points" throughout the series. A wider variation between the individual cells indicates trouble in the high or low cell.

Gravity above 1,200 indicates that the cells are more than half charged. Gravity below 1,150 indicates that the cells are completely discharged. Maximum gravity at full charge is 1,280 to 1,300.

Evaporation. In the course of time there is a loss of solution due to evaporation and to the spray passing through the vent holes. In all cases the level of the electrolyte should be kept above the tops of the plates since a low level causes sulphation of that part of the plate that comes into contact with the air and also causes a reduction in the current capacity of the cell.

Never pour pure acid or a strong solution of acid into a cell in renewing the electrolyte. The acid never mixes with the solution immediately and therefore soon attacks the active material on the plates. Renew all of the solution at the proper specific gravity after thoroughly mixing in another vessel, and allow it to cool thoroughly before pouring it into the cells. If it is desired to use the old solu-

229

tion, remove it from the cells before adding the acid, mix thoroughly with the new, and cool before pouring back.

If any particular cell regularly requires more water than the others, a leaky cell is indicated. Repair or renew the cell immediately. Sometimes a defective vent will cause excessive waste of electrolyte.

Low Gravity Cells. When the gravity of any one cell is lower than the others, say by more than 25 points, and if successive readings show that this difference is increasing, the cell is not in good order. If there is no leak in the battery jar, the low gravity generally indicates that there is an internal short circuit due to sediment, shedding, or buckled plates, and this trouble should be corrected immediately only by an expert battery man. Continued short circuits will eventually destroy the cell for the reason that they act in the same way as a continuous external load.

Standing Idle. A battery which is to stand idle for any length of time should first be fully charged as a protection against current leakage. It is not wise to permit a battery to stand idle for more than six months with the electrolyte in the cells. A battery not in active service should receive a freshening charge at least once in every two months. It should be given a thorough charge after an idle period before being replaced in service. After standing idle for two months or more it should be charged at one-half the normal rate to the maximum gravity.

Before leaving the battery for a prolonged idle period, disconnect the wires leading to the various circuits so that it will not lose its charge through any slight leaks in the wiring of the car. In cold weather it is best to store the battery in a heated room, for the electrolyte freezes at 20 degrees below zero.

When the batteries are to be left out of service for more than six months, proceed as follows: Discharge at the rate of about one ampere until the lamps burn with a dull red glow, empty out the acid and wash cell out thoroughly with clean water. Fill battery up with distilled water and discharge until there is no further current. Repeat the operation once more, and then fill up to the vent tubes with distilled water. The battery can now be set away for an indefinite period, the battery being visited occasionally to make up the water lost by evaporation.

Another method is as follows: Fully charge battery, empty acid from cells, clean out thoroughly, fill with distilled water, and let stand for a few moments. Empty cells, and repeat the operation three times, about one day between operations. Drain battery dry and set away in dry place.

Voltage Too High. Never allow the voltage to rise above 2.65

volts. A high voltage denotes that the density of the electrolyte is too high and therefore should be diluted with water.

Electrolyte or Solution. Only the purest of acid and distilled water should be used in making up the battery solution. Never use the commercial sulphuric acid commonly sold at drug stores as this contains traces of iron, platinum and lead which are highly injurious to the cell. Mix the electrolyte only in glass or porcelain vessels, as the acid will dissolve enough of a metal container to work havoc with the plates. Never stir a solution with an iron rod.

The electrolyte consists of one part of chemically pure sulphuric acid to four parts of distilled water. Pour the acid into the water, never the water into the acid. Severe explosion is likely to take place under the latter condition. Rain water can be used in an emergency, but should not be left in the cells longer than absolutely necessary.

Loss of Capacity. Loss of capacity may be caused by the active material shedding from the plates, by internal short circuits, by sediment in the bottom of the cells, by broken separators, by excessive temperatures or by the shrinkage of the active material.

Open Vents. When charging a battery be sure that the vent caps are fully open, for the generation of gas due to the charging process may burst the cell if not liberated.

Full Charge. A cell may be considered fully charged when, with the rate of current flow specified by the makers, all cells are gassing freely (bubbling) and evenly and the gravity shows no increase for one hour. The gravity should never be lower than 1,150 before charging. Six-volt batteries cannot be overcharged if the voltage of the generator is below 7 volts.

Care of Terminals. Keep the battery terminals thoroughly greased to prevent corrosion and the breaking of the conductor wires. The acid spray carried through the vent caps not only destroys the terminals but is also likely to carry the dissolved metal back into the battery solution and ruin the plates.

Charging. Start charging as soon as the voltage drops below 1.8, taking care not to exceed the charging rate specified by the maker of the battery. Be sure that the vents are open before closing the charging switch. When connecting the charging current to the batteries be sure that the positive pole of the line is connected to the positive pole of the battery so that the current flows "backwards" through the battery or in a direction opposite to that given by the cells when discharging. The positive wire is marked "POS" and is generally painted red.

DYNAMO AND MOTOR TROUBLES.

While there is comparatively little trouble with the motor or dynamo end of the self-starting system, all of the moving parts will wear in the course of time and will require attention. The majority of the diseases to which the dynamo is heir can be cured by the owner if he will take the trouble to systematically observe the symptoms of his patient and proceed carefully in accordance with the following instructions.

Primarily, the principal indications of trouble are sparking at the brushes and commutator, heating of the armature or field, noise, failure to generate in the case of a dynamo, or failure to start in the case of the motor. Before proceeding with dynamo or motor tests be sure that the battery and the external wiring are not at fault and that the switching and regulating systems are working freely. Wire is usually the part of the circuit that is the least protected from the effects of moisture and abrasion and therefore is the part to be first placed under suspicion. Even when protected with a metal conduit the insulation of the wire often deriorates before trouble is experienced with the battery, motor or dynamo.

Remember that moisture and oil are two of the greatest enemies of insulation and therefore of the electric system as a whole, and do not fail at all times to protect the windings and wiring if the most important point has been neglected by the maker of the machine.

SPARKING AT THE BRUSHES.

Remove hand-hole plates or doors above the commutator and examine the parts when the machine is generating or when using a maximum amount of current. Note whether the connections are tight, commutator is rough or cut, or whether the brushes are pressing with the proper tension on the surface of the commutator. Clean all of the current carrying surfaces before proceeding further with the inspection.

Poor Contact. Clean commutator thoroughly with gasoline after removing brushes. Soak brushes thoroughly in gasoline and carefully scrape off sharp corners on the bearing surfaces. If brushes are of carbon, alcohol is better than gasoline or benzine.

See that the bearing surface of the brushes is perfectly smooth and glossy and that the brush bears evenly on the commutator throughout its entire width. Slightly bevel the trailing and leading edges so that there is no danger of the corners scratching the commutator. If there are low spots in the bearing face of the brush there will not be sufficient area for the conduction of the current. Grind them to a uniform, even, concave surface with a piece of sand paper mounted on a block curved approximately to the outline of the commutator.

Tension Springs. The tension springs feeding the brush often slacken, lose their temper, or slip from their moorings so that the brush is not held on the commutator with sufficient pressure. This allows the brushes to joggle and hence to spark. Tighten the spring so that it gives a firm bearing yet without excessive friction on the commutator. Do not make it any heavier than necessary to stop the sparking. The slot in the brush holder in which the brush slides is often either too large or too small for the brush. If too large the brush will joggle, if too small or tight, the brush will bind so that the spring will not be able to feed it with the proper pressure on the commutator. See that the connection between the brush and the brush holder is electrically perfect so that current can enter the brush properly. See that the brush holder allows the front edge of the brush to bear exactly parallel with the edges of the commutator bars. A brush worn too short may release the spring tension.

Chattering. A chattering noise while the machine is running, with an excessive vibration in the brush, may be caused either by the brush or the commutator. If in the brush, it is generally caused by a loose brush or loose brush holder, by a very hard carbon brush, or by insufficient spring pressure. If the commutator is lubricated by a slight amount of vaseline applied with a rag it will often be found to stop the noise. Never use oil, especially when applied with an oil can. If it must be used slightly dampen a soft rag and apply while the machine is running. Paraffin wax applied with slight pressure is an excellent lubricant.

Grooves. Grooves either in the brushes or commutator are often produced by grit or hard spots in the brushes. Use the best grade of carbon (Soft), or the proper grade of copper for the brushes.

Uneven Spacing. The tips of the brushes should bear on the commutator at points diametrically opposite to one another. The truth of this position can be determined by counting the number of commutator bars that lay on either side of the brush. This number should in all cases be equal. Uneven spacing is not a common trouble with automobile generators as the brush holders are generally fastened in a fixed position.

Dirty Commutator. Clean with gasoline or benzine. See that there are no copper filings or dust on the insulation at the ends of the bars or between the lugs at the point where the armature wires are fastened to the commutator bars. If oil is thrown from the bearings upon the commutator a shield should be installed to protect the insulation.

Grooves. If the surface of the commutator appears to have a series of fine cuts or grooves grind down with a piece of very fine sand or glass paper mounted on block. The face to which the sand paper is applied should be curved approximately to the radius of the commutator so that the cutting will be even. If the grooves are very deep it is easier to remove the armature and turn the commutator down on a lathe. This also insures a perfectly cylindrical commutator. While the grooves are generally caused in the first place by the brushes, the grooves on the commutator react on the brushes causing rapid wear, sparking, and brush chatter.

Never use emery in grinding down, nor carborundum as these materials have the property of sticking into the copper bars after the operation. Thoroughly clean off all copper dust and sand particles caused by grinding.

When the cutting of grooves is persistent even after having changed the brushes, look to the tension springs of the brush holder, for these may be causing too much pressure on the brushes. The armature or commutator should have a small amount of end play in the bearings to prevent the brushes from tracking continually in one path.

Burned Bars. Black burned spots on the commutator may be caused by chattering brushes, insufficient spring tension, or by open circuit or grounds in the armature windings. A ground in a commutator bar or a short circuit between two adjacent bars will also cause a burned spot. Burning due to brush trouble is generally spread uniformly over the entire surface of the commutator, and in the path of the brushes. Burning due to armature winding troubles occurs at only one or two bars, the principal burning being at the trailing edge of the bar, or at the point where the last live bar leaves the brush. If this is allowed to continue, the insulation will be eventually burned out between the bars, and the entire commutator grounded. Burned bars should be sand papered down as described in (8), and the insulation between the bars should be repaired by digging out the burned mica and replacing the burned spot with fresh mica mixed in shellac. Dig out only that portion that is burned.

Out of Round. When the commutator wears out of round, it causes the brushes to vibrate rapidly up and down, thus producing burns and unnecessary vibration. It also unbalances the rotating mass to a certain extent and is a factor in producing bear-wear and uncomfortable noise. The only remedy is to turn the commutator down in a lathe.

High Bar. Owing to the loosening of the end retaining rings or to shrinkage of the commutator insulation, a bar often rises above the adjacent bars, causing burns, noise and unnecessary brush wear. When this occurs the armature should be removed and sent to a competent armature repair man for this is a trouble that cannot usually be made satisfactorily by the amateur. If the repair must be performed by the garage man or owner it can sometimes be done by tapping down the bar with a piece of wood and a mallet, tightening the end nuts that hold the bars on the commutator spider, and then turning the whole down on a lathe. Care should be taken not to injure the insulation at the ends of the bars.

Low Bar. Symptoms and treatment the same as for high bar.

Weak Magnetic Field. A weak magnetic field will cause sparking, especially in the case of a motor, since this produces a change in the commutating point on which the brushes should rest. A weak field may be caused by a poor external connection, which in turn causes undue resistance in the field circuit, or by a short circuit, broken field wire, or dampness in the windings. A grounded or wet wire leading to the field winding will also cause a weak field. In a generator, a weak field will cause a fall in voltage when the fault is in the shunt field, and a maintained high voltage when the fault is in the series field, the lighting generators all being of the differentially wound compound type.

A weak motor field causes a reduced torque with a load, and overspeeding with the load removed. The only remedy is to replace the wires leading to the fields externally, or to rewind the field coils if the trouble is internal. If a switching device is included in the field circuit see that it is making proper contact and that all connections are tight.

Excessive Loads. When the motor or generator are carrying an excessive current, sparking is almost certain to appear. An excessive current in a generator may be caused by a defective current regulating device which causes the voltage to rise too far above the voltage of the battery, or it may be caused by a defective field winding in cases where no regulating device is used. The faulty field results in a high charging voltage. Sparking sometimes occurs when the voltage of the battery drops to a very low point due to large drafts of current by the battery.

An excessive motor load may be caused by a cold stiff engine, by leaving the clutch in when starting the engine, by attempting to drive the car with the starting motor, or by having the ignition advanced too far. Hot bearings in either the starting motor or the automobile engine will produce the same results.

Short circuits and grounds in the wires leading to the battery and in the external circuit are often the cause of abnormal currents through the generator. Test out the wiring on the chassis of the car.

Friction. Try out the engine, starting motor, and generator for bearing and gearing binding and friction. Be sure that the pistons are not stuck, and that no parts are frozen together during cold spells. It should be remembered that an engine always cranks heavier in cold than in warm weather owing to stiff oil.

Short Circuited Armature Coils. Remove all filings, copper dust, solder and other metallic connections behind the commutator bars. See that the clamping rings are perfectly insulated from the bars, and that no bridge exists through dust or solder. Test for grounds and see that the brush holders are perfectly insulated.

Broken Armature Coils. Examine commutator bars at the point where the connection is made to the wires leading from the armature windings. If a loose or broken wire is found at this point it should be replaced. If the coil is broken in the armature, rewinding is the only sure remedy, but it can be temporarily repaired by bridging across the broken coil on the ends of the commutator lugs. Connect the bar leading to the broken coil to the next bar on either side across the mica. In cases where the brush holders can be shifted or the brushes turned in the holder, the brush should be turned so that it will bridge across two bars simultaneously. Temporarily, the two bars lying on either side of the bar that connects with the broken coil can be connected together with a jumper, and the ends of the broken wire removed from the lug.

In adopting any repair that shorts or cuts out an armature coil, remember that the voltage and speed will be varied according to the number of active coils cut out. Prompt rewinding is the only safe course.

HEATING.

Overload. Too many lights, or too many amperes supplied to or taken from the machine may cause heating. Defects in the cut-out mechanism, or in the field windings will cause overloads. Running the car on the starting motor will cause excessive heat, as will a stiff cold engine.

Defective Cut-Out. A defective cut-out will allow the current from the battery to rush back through the generator when the voltage of the generator falls below the voltage of the battery.

Short Circuit (External). A short circuit in the wires leading to the various parts of the car will put a heavy load on the generator and consequently will cause over-heating.

Moisture. Dry out the fields or armature by gentle heat either in an oven or by sending a small current through the coils. When it is known that the motor has been wet it should be dried before running it, or before trouble has developed.

Short Circuit (Internal). A short circuit due to dirt at the commutator, or to abraded insulation in the armature winding will cause heating. A ground in the armature windings will often have the same effect as a short circuit. In the field coil a short circuit or moisture will produce a weak field (See 13), causing trouble at the brushes and the commutator, or heating in the shunt fields of the generator.

www.ingramcontent.com/pod-product-compliance
Lightning Source LLC
Chambersburg PA
CBHW021922190326
41519CB00009B/878